ACT Subject Test

Mathematics

Student Practice Workbook

+ Two Full-Length ACT Math Tests

Math Notion

www.MathNotion.com

ACT Subject Test Mathematics

Published in the United State of America By

The Math Notion

Web: WWW.MathNotion.com

Email: info@Mathnotion.com

Copyright © 2021 by the Math Notion. All rights reserved. No part of this publication may be reproduced, stored in a retrieval system, or transmitted in any form or by any means, electronic, mechanical, photocopying, recording, scanning, or otherwise, except as permitted under Section 107 or 108 of the 1976 United States Copyright Ac, without permission of the author.

All inquiries should be addressed to the Math Notion.

ISBN: 978-1-63620-048-4

The Math Notion

Michael Smith has been a math instructor for over a decade now. He launched the Math Notion. Since 2006, we have devoted our time to both teaching and developing exceptional math learning materials. As a test prep company, we have worked with thousands of students. We have used the feedback of our students to develop a unique study program that can be used by students to drastically improve their math scores fast and effectively. We have more than a thousand Math learning books including:

- **SAT Math Prep**
- **PSAT Math Prep**
- **GRE Math Prep**
- **Accuplacer Math Prep**
- **Common Core Math Prep**
- **many Math Education Workbooks, Study Guides, Practice and Exercise Books**

As an experienced Math test preparation company, we have helped many students raise their standardized test scores—and attend the colleges of their dreams: We tutor online and in person, we teach students in large groups, and we provide training materials and textbooks through our website and through Amazon.

You can contact us via email at:

info@Mathnotion.com

Get the Targeted Practice You Need to Ace the ACT Math Test!

ACT Subject Test - Mathematics includes easy-to-follow instructions, helpful examples, and plenty of math practice problems to assist students to master each concept, brush up their problem-solving skills, and create confidence.

The ACT math practice book provides numerous opportunities to evaluate basic skills along with abundant remediation and intervention activities. It is a skill that permits you to quickly master intricate information and produce better leads in less time.

Students can boost their test-taking skills by taking the book's two practice ACT Math exams. All test questions answered and explained in detail.

Important Features of the ACT Math Book:

- A **complete review** of ACT math test topics,
- Over 2,500 practice problems covering all topics tested,
- The most important concepts you need to know,
- Clear and concise, easy-to-follow sections,
- Well designed for enhanced learning and interest,
- Hands-on experience with all question types
- **2 full-length practice tests** with detailed answer explanations
- Cost-Effective Pricing

Powerful math exercises to help you avoid traps and pacing yourself to beat the ACT test. Students will gain valuable experience and raise their confidence by taking math practice tests, learning about test structure, and gaining a deeper understanding of what is tested on the ACT Math. If ever there was a book to respond to the pressure to increase students' test scores, this is it.

WWW.MathNotion.COM

... So Much More Online!

✓ FREE Math Lessons

✓ More Math Learning Books!

✓ Mathematics Worksheets

✓ Online Math Tutors

For a PDF Version of This Book

Please Visit WWW.MathNotion.com

Contents

Chapter 1 :

Integers and Number Theory

Topics that you'll practice in this chapter:

- ✓ Rounding
- ✓ Whole Number Addition and Subtraction
- ✓ Whole Number Multiplication and Division
- ✓ Rounding and Estimates
- ✓ Adding and Subtracting Integers
- ✓ Multiplying and Dividing Integers
- ✓ Order of Operations
- ✓ Ordering Integers and Numbers
- ✓ Integers and Absolute Value
- ✓ Factoring Numbers
- ✓ Greatest Common Factor (GCF)
- ✓ Least Common Multiple (LCM)

"Wherever there is number, there is beauty." –*Proclus*

Rounding

✎ **Round each number to the nearest ten.**

1) 42 = ____   5) 19 = ____   9) 48 = ____

2) 88 = ____   6) 25 = ____   10) 81 = ____

3) 24 = ____   7) 93 = ____   11) 58 = ____

4) 57 = ____   8) 71 = ____   12) 87 = ____

✎ **Round each number to the nearest hundred.**

13) 198 = ____   17) 321 = ____   21) 580 = ____

14) 387 = ____   18) 433 = ____   22) 868 = ____

15) 816 = ____   19) 579 = ____   23) 480 = ____

16) 101 = ____   20) 825 = ____   24) 287 = ____

✎ **Round each number to the nearest thousand.**

25) 1,382 = ____   29) 9,099 = ____   33) 52,866 = ____

26) 3,420 = ____   30) 22,980 = ____   34) 85,190 = ____

27) 4,254 = ____   31) 45,188 = ____   35) 70,990 = ____

28) 6,861 = ____   32) 16,808 = ____   36) 26,869 = ____

Rounding and Estimates

✏ **Estimate the sum by rounding each number to the nearest ten.**

1) $13 + 22 = $ _____

2) $71 + 23 = $ _____

3) $61 + 58 = $ _____

4) $56 + 85 = $ _____

5) $368 + 249 = $ _____

6) $330 + 903 = $ _____

7) $471 + 293 = $ _____

8) $1,950 + 2,655 = $ _____

✏ **Estimate the product by rounding each number to the nearest ten.**

9) $32 \times 71 = $ _____

10) $12 \times 33 = $ _____

11) $31 \times 83 = $ _____

12) $19 \times 11 = $ _____

13) $42 \times 76 = $ _____

14) $63 \times 34 = $ _____

15) $19 \times 31 = $ _____

16) $59 \times 71 = $ _____

✏ **Estimate the sum or product by rounding each number to the nearest ten.**

17) $\begin{array}{r} 29 \\ \times\ 12 \\ \hline \end{array}$

18) $\begin{array}{r} 37 \\ \times\ 26 \\ \hline \end{array}$

19) $\begin{array}{r} 48 \\ +\ 82 \\ \hline \end{array}$

20) $\begin{array}{r} 65 \\ +44 \\ \hline \end{array}$

21) $\begin{array}{r} 37 \\ \times\ 14 \\ \hline \end{array}$

22) $\begin{array}{r} 71 \\ +\ 32 \\ \hline \end{array}$

Adding and Subtracting Integers

✎ **Find each sum.**

1) $14 + (-6) =$

2) $(-13) + (-20) =$

3) $5 + (-28) =$

4) $50 + (-12) =$

5) $(-7) + (-15) + 3 =$

6) $30 + (-14) + 8 =$

7) $40 + (-10) + (-14) + 17 =$

8) $(-15) + (-20) + 13 + 35 =$

9) $40 + (-20) + (38 - 29) =$

10) $28 + (-12) + (30 - 12) =$

✎ **Find each difference.**

11) $(-18) - (-7) =$

12) $25 - (-14) =$

13) $(-20) - 36 =$

14) $34 - (-19) =$

15) $51 - (30 - 21) =$

16) $17 - (5) - (-24) =$

17) $(35 + 20) - (-46) =$

18) $48 - 16 - (-8) =$

19) $62 - (28 + 17) - (-15) =$

20) $58 - (-23) - (-31) =$

21) $19 - (-8) - (-13) =$

22) $(19 - 24) - (-14) =$

23) $27 - 33 - (-21) =$

24) $58 - (32 + 24) - (-9) =$

25) $36 - (-30) + (-17) =$

26) $27 - (-42) + (-31) =$

Multiplying and Dividing Integers

✏ **Find each product.**

1) $(-9) \times (-5) =$

2) $(-3) \times 9 =$

3) $8 \times (-12) =$

4) $(-7) \times (-20) =$

5) $(-3) \times (-5) \times 6 =$

6) $(14 - 3) \times (-8) =$

7) $12 \times (-9) \times (-3) =$

8) $(140 + 10) \times (-2) =$

9) $10 \times (-12 + 8) \times 3 =$

10) $(-8) \times (-5) \times (-10) =$

✏ **Find each quotient.**

11) $42 \div (-7) =$

12) $(-48) \div (-6) =$

13) $(-40) \div (-8) =$

14) $54 \div (-2) =$

15) $152 \div 19 =$

16) $(-144) \div (-12) =$

17) $180 \div (-10) =$

18) $(-312) \div (-12) =$

19) $221 \div (-13) =$

20) $(-126) \div (6) =$

21) $(-161) \div (-7) =$

22) $-266 \div (-14) =$

23) $(-120) \div (-4) =$

24) $270 \div (-18) =$

25) $(-208) \div (-8) =$

26) $(135) \div (-15) =$

Order of Operations

✎ **Evaluate each expression.**

1) $7 + (5 \times 4) =$

2) $14 - (3 \times 6) =$

3) $(19 \times 4) + 16 =$

4) $(16 - 7) - (8 \times 2) =$

5) $27 + (18 \div 3) =$

6) $(18 \times 8) \div 6 =$

7) $(32 \div 4) \times (-2) =$

8) $(9 \times 4) + (32 - 18) =$

9) $24 + (4 \times 3) + 7 =$

10) $(36 \times 3) \div (2 + 2) =$

11) $(-7) + (12 \times 3) + 11 =$

12) $(8 \times 5) - (24 \div 6) =$

13) $(7 \times 6 \div 3) - (12 + 9) =$

14) $(13 + 5 - 14) \times 3 - 2 =$

15) $(20 - 14 + 30) \times (64 \div 4) =$

16) $32 + \big(28 - (36 \div 9)\big) =$

17) $(7 + 6 - 4 - 7) + (15 \div 5) =$

18) $(85 - 20) + (20 - 18 + 7) =$

19) $(20 \times 2) + (14 \times 3) - 22 =$

20) $18 + 5 - (30 \times 3) + 20 =$

Ordering Integers and Numbers

✎ **Order each set of integers from least to greatest.**

1) $8, -10, -5, -3, 4$ ___, ___, ___, ___, ___, ___

2) $-10, -18, 6, 14, 27$ ___, ___, ___, ___, ___, ___

3) $15, -8, -21, 21, -23$ ___, ___, ___, ___, ___, ___

4) $-14, -40, 23, -12, 47$ ___, ___, ___, ___, ___, ___

5) $59, -54, 32, -57, 36$ ___, ___, ___, ___, ___, ___

6) $68, 26, -19, 47, -34$ ___, ___, ___, ___, ___, ___

✎ **Order each set of integers from greatest to least.**

7) $18, 36, -16, -18, -10$ ___, ___, ___, ___, ___, ___

8) $27, 34, -12, -24, 94$ ___, ___, ___, ___, ___, ___

9) $50, -21, -13, 42, -2$ ___, ___, ___, ___, ___, ___

10) $37, 46, -20, -16, 86$ ___, ___, ___, ___, ___, ___

11) $-18, 88, -26, -59, 75$ ___, ___, ___, ___, ___, ___

12) $-65, -30, -25, 3, 14$ ___, ___, ___, ___, ___, ___

Integers and Absolute Value

✎ **Write absolute value of each number.**

1) $|-2| =$

2) $|-27| =$

3) $|-20| =$

4) $|14| =$

5) $|6| =$

6) $|-55| =$

7) $|16| =$

8) $|2| =$

9) $|54| =$

10) $|-4| =$

11) $|-11|$

12) $|88| =$

13) $|0| =$

14) $|79| =$

15) $|-32| =$

16) $|-17| =$

17) $|42| =$

18) $|-46| =$

19) $|1| =$

20) $|-40| =$

✎ **Evaluate the value.**

21) $|-5| - \frac{|-21|}{7} =$

22) $14 - |3 - 15| - |-4| =$

23) $\frac{|-32|}{4} \times |-4| =$

24) $\frac{|7 \times (-3)|}{7} \times \frac{|-19|}{3} =$

25) $|4 \times (-5)| + \frac{|-40|}{5} =$

26) $\frac{|-45|}{9} \times \frac{|-24|}{12} =$

27) $|-12 + 8| \times \frac{|-7 \times 7|}{7} =$

28) $\frac{|-11 \times 2|}{4} \times |-16| =$

Factoring Numbers

✎ **List all positive factors of each number.**

1) 9

2) 16

3) 24

4) 30

5) 26

6) 46

7) 20

8) 68

9) 28

10) 98

11) 14

12) 54

13) 55

14) 18

15) 63

16) 34

17) 50

18) 62

19) 95

20) 64

21) 70

22) 45

23) 22

24) 65

Greatest Common Factor

✏ **Find the GCF for each number pair.**

1) 6, 2	9) 12, 18	17) 42, 14
2) 4, 5	10) 4, 36	18) 16, 40
3) 3, 12	11) 6, 10	19) 9, 2, 3
4) 7, 3	12) 28, 52	20) 5, 15, 10
5) 5, 10	13) 25, 10	21) 7, 9, 2
6) 8, 48	14) 22, 24	22) 16, 64
7) 6, 18	15) 9, 54	23) 30, 48
8) 9, 15	16) 8, 54	24) 36, 63

Least Common Multiple

✎ **Find the LCM for each number pair.**

1) 6, 9

2) 15, 45

3) 16, 40

4) 12, 36

5) 18, 27

6) 14, 42

7) 6, 30

8) 8, 56

9) 7, 21

10) 8, 20

11) 15, 25

12) 7, 9

13) 4, 11

14) 8, 28

15) 28, 56

16) 40, 50

17) 12, 13

18) 22, 11

19) 36, 20

20) 15, 35

21) 18, 81

22) 30, 54

23) 18, 45

24) 75, 25

ACT Subject Test – Mathematics

Answers of Worksheets

Rounding

1) 40	10) 80	19) 600	28) 7,000
2) 90	11) 60	20) 800	29) 9,000
3) 20	12) 90	21) 600	30) 23,000
4) 60	13) 200	22) 900	31) 45,000
5) 20	14) 400	23) 500	32) 17,000
6) 30	15) 800	24) 300	33) 53,000
7) 90	16) 100	25) 1,000	34) 85,000
8) 70	17) 300	26) 3,000	35) 71,000
9) 50	18) 400	27) 4,000	36) 27,000

Rounding and Estimates

1) 30	7) 760	13) 3,200	19) 130
2) 90	8) 4,610	14) 1,800	20) 110
3) 120	9) 2,100	15) 600	21) 400
4) 150	10) 300	16) 4,200	22) 100
5) 620	11) 2,400	17) 300	
6) 1,230	12) 200	18) 1,200	

Adding and Subtracting Integers

1) 8	8) 13	15) 42	22) 9
2) −33	9) 29	16) 36	23) 15
3) −23	10) 34	17) 101	24) 11
4) 38	11) −11	18) 40	25) 49
5) −19	12) 39	19) 32	26) 38
6) 24	13) −56	20) 112	
7) 33	14) 53	21) 40	

Multiplying and Dividing Integers

1) 45	6) −88	11) −6	16) 12
2) −27	7) 324	12) 8	17) −18
3) −96	8) −300	13) 5	18) 26
4) 140	9) −120	14) −27	19) −17
5) 90	10) −400	15) 8	20) −21

21) 23   23) 30   25) 26

22) 19   24) −15   26) −9

Order of Operations

1) 27   6) 24   11) 40   16) 56

2) −4   7) −16   12) 36   17) 5

3) 92   8) 50   13) −7   18) 74

4) −7   9) 43   14) 10   19) 60

5) 33   10) 27   15) 576   20) −47

Ordering Integers and Numbers

1) −10, −5, −3, 4, 8   7) 36, 18, −10, −16, −18

2) −18, −10, 6, 14, 27   8) 94, 34, 27, −12, −24

3) −23, −21, −8, 15, 21   9) 50, 42, −2, −13, −21

4) −40, −14, −12, 23, 47   10) 86, 46, 37, −16, −20

5) −57, −54, 32, 36, 59   11) 88, 75, −18, −26, −59

6) −34, −19, 26, 47, 68   12) 14, 3, −25, −30, −65

Integers and Absolute Value

1) 2   8) 2   15) 32   22) −2

2) 27   9) 54   16) 17   23) 32

3) 20   10) 4   17) 42   24) 19

4) 14   11) 11   18) 46   25) 28

5) 6   12) 88   19) 1   26) 10

6) 55   13) 0   20) 40   27) 28

7) 16   14) 79   21) 2   28) 88

Factoring Numbers

1) 1, 3, 9   9) 1, 2, 4, 7, 14, 28   17) 1, 2, 5, 10, 25, 50

2) 1, 2, 4, 8, 16   10) 1, 2, 7, 14, 49, 98   18) 1, 2, 31, 62

3) 1, 2, 3, 4, 6, 8, 12, 24   11) 1, 2, 7, 14   19) 1, 5, 19, 95

4) 1, 2, 3, 5, 6, 10, 15, 30   12) 1, 2, 3, 6, 9, 18, 27, 54   20) 1, 2, 4, 8, 16, 32, 64

5) 1, 2, 13, 26   13) 1, 5, 11, 55   21) 1, 2, 5, 7, 10, 14, 35, 70

6) 1, 2, 23, 46   14) 1, 2, 3, 6, 9, 18   22) 1, 3, 5, 9, 15, 45

7) 1, 2, 4, 5, 10, 20 15) 1, 3, 7, 9, 21, 63   23) 1, 2, 11, 22

8) 1, 2, 4, 17, 34, 68 16) 1, 2, 17, 34   24) 1, 5, 13, 65

Greatest Common Factor

1) 2	7) 6	13) 5	19) 1
2) 1	8) 3	14) 2	20) 5
3) 3	9) 6	15) 9	21) 1
4) 1	10) 4	16) 2	22) 16
5) 5	11) 2	17) 14	23) 6
6) 8	12) 4	18) 8	24) 9

Least Common Multiple

1) 18	7) 30	13) 44	19) 180
2) 45	8) 56	14) 56	20) 105
3) 80	9) 21	15) 56	21) 162
4) 36	10) 40	16) 200	22) 270
5) 54	11) 75	17) 156	23) 90
6) 42	12) 63	18) 22	24) 75

Chapter 2 :
Fractions and Decimals

Topics that you'll practice in this chapter:

✓ Simplifying Fractions

✓ Adding and Subtracting Fractions

✓ Multiplying and Dividing Fractions

✓ Adding and Subtract Mixed Numbers

✓ Multiplying and Dividing Mixed Numbers

✓ Adding and Subtracting Decimals

✓ Multiplying and Dividing Decimals

✓ Comparing Decimals

✓ Rounding Decimals

"A Man is like a fraction whose numerator is what he is and whose denominator is what he thinks of himself. The larger the denominator, the smaller the fraction." –Tolstoy

Simplifying Fractions

✎ **Simplify each fraction to its lowest terms.**

1) $\frac{5}{10} =$ 8) $\frac{15}{60} =$ 15) $\frac{35}{105} =$

2) $\frac{28}{35} =$ 9) $\frac{80}{160} =$ 16) $\frac{25}{70} =$

3) $\frac{27}{36} =$ 10) $\frac{55}{77} =$ 17) $\frac{80}{280} =$

4) $\frac{40}{80} =$ 11) $\frac{28}{112} =$ 18) $\frac{12}{81} =$

5) $\frac{14}{56} =$ 12) $\frac{32}{64} =$ 19) $\frac{36}{186} =$

6) $\frac{32}{48} =$ 13) $\frac{63}{72} =$ 20) $\frac{240}{540} =$

7) $\frac{52}{65} =$ 14) $\frac{81}{90} =$ 21) $\frac{70}{560} =$

✎ **Find the answer for each problem.**

22) Which of the following fractions equal to $\frac{3}{4}$? _____

     A. $\frac{60}{90}$ B. $\frac{43}{104}$ C. $\frac{48}{64}$ D. $\frac{150}{300}$

23) Which of the following fractions equal to $\frac{5}{8}$? _____

     A. $\frac{125}{200}$ B. $\frac{115}{200}$ C. $\frac{50}{100}$ D. $\frac{30}{90}$

24) Which of the following fractions equal to $\frac{3}{7}$? _____

     A. $\frac{58}{116}$ B. $\frac{54}{126}$ C. $\frac{270}{167}$ D. $\frac{42}{63}$

Adding and Subtracting Fractions

✎ **Find the sum.**

1) $\frac{5}{9} + \frac{4}{9} =$

2) $\frac{1}{2} + \frac{1}{7} =$

3) $\frac{3}{8} + \frac{1}{4} =$

4) $\frac{3}{5} + \frac{1}{2} =$

5) $\frac{1}{4} + \frac{3}{5} =$

6) $\frac{7}{8} + \frac{3}{8} =$

7) $\frac{1}{2} + \frac{7}{10} =$

8) $\frac{2}{5} + \frac{2}{3} =$

9) $\frac{5}{7} + \frac{2}{3} =$

10) $\frac{7}{12} + \frac{3}{4} =$

11) $\frac{5}{6} + \frac{2}{5} =$

12) $\frac{1}{12} + \frac{2}{3} =$

✎ **Find the difference.**

13) $\frac{1}{3} - \frac{1}{6} =$

14) $\frac{3}{4} - \frac{1}{8} =$

15) $\frac{1}{2} - \frac{1}{3} =$

16) $\frac{1}{4} - \frac{1}{5} =$

17) $\frac{5}{8} - \frac{2}{3} =$

18) $\frac{1}{4} - \frac{1}{7} =$

19) $\frac{5}{6} - \frac{1}{9} =$

20) $\frac{3}{4} - \frac{1}{6} =$

21) $\frac{7}{8} - \frac{1}{12} =$

22) $\frac{8}{15} - \frac{3}{5} =$

23) $\frac{3}{12} - \frac{1}{14} =$

24) $\frac{10}{13} - \frac{7}{26} =$

25) $\frac{6}{7} - \frac{3}{4} =$

26) $\frac{4}{5} - \frac{1}{8} =$

27) $\frac{4}{7} - \frac{2}{35} =$

28) $\frac{9}{16} - \frac{2}{8} =$

29) $\frac{8}{9} - \frac{7}{18} =$

30) $\frac{1}{2} - \frac{4}{9} =$

Multiplying and Dividing Fractions

✍ **Find the value of each expression in lowest terms.**

1) $\frac{1}{5} \times \frac{15}{5} =$

2) $\frac{9}{12} \times \frac{4}{9} =$

3) $\frac{1}{16} \times \frac{8}{10} =$

4) $\frac{1}{24} \times \frac{8}{10} =$

5) $\frac{1}{5} \times \frac{1}{4} =$

6) $\frac{7}{9} \times \frac{1}{7} =$

7) $\frac{6}{7} \times \frac{1}{3} =$

8) $\frac{2}{8} \times \frac{2}{8} =$

9) $\frac{5}{8} \times \frac{3}{5} =$

10) $\frac{4}{7} \times \frac{1}{8} =$

11) $\frac{7}{15} \times \frac{5}{7} =$

12) $\frac{3}{10} \times \frac{5}{9} =$

✍ **Find the value of each expression in lowest terms.**

13) $\frac{1}{4} \div \frac{1}{8} =$

14) $\frac{1}{10} \div \frac{1}{5} =$

15) $\frac{3}{4} \div \frac{1}{5} =$

16) $\frac{1}{3} \div \frac{5}{6} =$

17) $\frac{1}{7} \div \frac{8}{42} =$

18) $\frac{3}{4} \div \frac{1}{6} =$

19) $\frac{2}{7} \div \frac{7}{13} =$

20) $\frac{1}{24} \div \frac{3}{16} =$

21) $\frac{7}{12} \div \frac{5}{6} =$

22) $\frac{22}{18} \div \frac{11}{9} =$

23) $\frac{9}{35} \div \frac{3}{7} =$

24) $\frac{2}{7} \div \frac{8}{21} =$

25) $\frac{1}{9} \div \frac{2}{5} =$

26) $\frac{5}{12} \div \frac{3}{5} =$

27) $\frac{3}{20} \div \frac{1}{6} =$

28) $\frac{8}{20} \div \frac{3}{4} =$

29) $\frac{5}{6} \div \frac{2}{9} =$

30) $\frac{5}{11} \div \frac{3}{4} =$

Adding and Subtracting Mixed Numbers

✍ **Find the sum.**

1) $3\frac{1}{3} + 2\frac{1}{6} =$

2) $4\frac{1}{2} + 3\frac{1}{2} =$

3) $3\frac{3}{8} + 1\frac{1}{8} =$

4) $2\frac{1}{4} + 2\frac{1}{3} =$

5) $3\frac{5}{6} + 2\frac{7}{12} =$

6) $5\frac{4}{15} + 3\frac{3}{5} =$

7) $2\frac{1}{3} + 4\frac{3}{7} =$

8) $3\frac{1}{2} + 4\frac{2}{5} =$

9) $5\frac{2}{5} + 6\frac{3}{7} =$

10) $8\frac{5}{16} + 6\frac{1}{12} =$

✍ **Find the difference.**

11) $3\frac{1}{4} - 1\frac{3}{4} =$

12) $6\frac{3}{5} - 4\frac{2}{5} =$

13) $4\frac{1}{3} - 3\frac{1}{9} =$

14) $7\frac{1}{7} - 5\frac{1}{2} =$

15) $5\frac{1}{3} - 2\frac{1}{12} =$

16) $8\frac{1}{5} - 4\frac{1}{3} =$

17) $9\frac{1}{4} - 6\frac{1}{8} =$

18) $11\frac{7}{15} - 8\frac{3}{5} =$

19) $14\frac{5}{6} - 11\frac{3}{5} =$

20) $18\frac{2}{7} - 14\frac{1}{5} =$

21) $9\frac{1}{3} - 4\frac{1}{4} =$

22) $6\frac{1}{8} - 4\frac{1}{16} =$

23) $19\frac{3}{8} - 15\frac{1}{3} =$

24) $11\frac{1}{9} - 8\frac{1}{8} =$

25) $17\frac{1}{7} - 11\frac{1}{5} =$

26) $16\frac{2}{9} - 9\frac{5}{7} =$

Multiplying and Dividing Mixed Numbers

✏ **Find the product.**

1) $5\frac{1}{2} \times 2\frac{1}{4} =$

2) $5\frac{1}{3} \times 4\frac{1}{3} =$

3) $5\frac{3}{4} \times 6\frac{1}{4} =$

4) $3\frac{1}{3} \times 2\frac{3}{5} =$

5) $4\frac{8}{10} \times 1\frac{1}{24} =$

6) $6\frac{2}{7} \times 1\frac{1}{11} =$

7) $8\frac{2}{3} \times 3\frac{1}{2} =$

8) $3\frac{4}{7} \times 2\frac{1}{5} =$

9) $5\frac{2}{8} \times 4\frac{1}{6} =$

10) $7\frac{3}{3} \times 1\frac{3}{8} =$

✏ **Find the quotient.**

11) $2\frac{2}{5} \div 4\frac{1}{5} =$

12) $4\frac{1}{6} \div 3\frac{1}{3} =$

13) $6\frac{1}{3} \div 1\frac{1}{2} =$

14) $7\frac{1}{10} \div 2\frac{2}{5} =$

15) $3\frac{1}{3} \div 1\frac{1}{9} =$

16) $1\frac{1}{10} \div 4\frac{1}{2} =$

17) $1\frac{3}{16} \div 5\frac{1}{4} =$

18) $4\frac{1}{3} \div 4\frac{3}{4} =$

19) $9\frac{1}{3} \div 2\frac{1}{4} =$

20) $15\frac{1}{3} \div 5\frac{1}{2} =$

21) $4\frac{1}{6} \div 1\frac{1}{5} =$

22) $1\frac{1}{18} \div 1\frac{2}{9} =$

23) $4\frac{2}{7} \div 1\frac{3}{10} =$

24) $7\frac{1}{3} \div 2\frac{2}{11} =$

25) $8\frac{2}{5} \div 1\frac{1}{6} =$

26) $9\frac{1}{3} \div 2\frac{1}{7} =$

Adding and Subtracting Decimals

✎ **Add and subtract decimals.**

$$1) \begin{array}{r} 35.19 \\ -\ 24.28 \\ \hline \end{array}$$

$$4) \begin{array}{r} 38.72 \\ -\ 21.68 \\ \hline \end{array}$$

$$7) \begin{array}{r} 86.09 \\ -\ 35.14 \\ \hline \end{array}$$

$$2) \begin{array}{r} 34.29 \\ +\ 42.58 \\ \hline \end{array}$$

$$5) \begin{array}{r} 57.39 \\ +\ 26.54 \\ \hline \end{array}$$

$$8) \begin{array}{r} 54.51 \\ +\ 32.66 \\ \hline \end{array}$$

$$3) \begin{array}{r} 61.20 \\ +\ 33.75 \\ \hline \end{array}$$

$$6) \begin{array}{r} 70.24 \\ -\ 42.35 \\ \hline \end{array}$$

$$9) \begin{array}{r} 114.21 \\ -\ 88.69 \\ \hline \end{array}$$

✎ **Find the missing number.**

10) ___ $+ 2.8 = 5.4$

11) $4.1 +$ ___ $= 5.88$

12) $6.45 +$ ___ $= 8$

13) $7.25 -$ ___ $= 3.40$

14) ___ $- 2.35 = 4.25$

15) ___ $- 19.85 = 6.54$

16) $22.15 +$ ___ $= 28.95$

17) ___ $- 37.16 = 9.42$

18) ___ $+ 24.50 = 34.19$

19) $72.40 +$ ___ $= 125.20$

Multiplying and Dividing Decimals

✍ **Find the product.**

1) $0.5 \times 0.6 =$

2) $3.3 \times 0.4 =$

3) $1.28 \times 0.5 =$

4) $0.35 \times 0.6 =$

5) $1.85 \times 0.6 =$

6) $0.24 \times 0.5 =$

7) $5.25 \times 1.4 =$

8) $18.5 \times 4.6 =$

9) $15.4 \times 6.8 =$

10) $19.5 \times 2.6 =$

11) $32.2 \times 1.5 =$

12) $78.4 \times 4.5 =$

✍ **Find the quotient.**

13) $1.85 \div 10 =$

14) $74.6 \div 100 =$

15) $3.6 \div 3 =$

16) $9.6 \div 0.4 =$

17) $15.5 \div 0.5 =$

18) $32.8 \div 0.2 =$

19) $22.15 \div 1,000 =$

20) $53.55 \div 0.7 =$

21) $322.2 \div 0.2 =$

22) $50.67 \div 0.18 =$

23) $77.4 \div 0.8 =$

24) $27.93 \div 0.03 =$

Comparing Decimals

✎ **Write the correct comparison symbol (>, < or =).**

1) 0.70 ☐ 0.070

2) 0.049 ☐ 0.49

3) 5.090 ☐ 5.09

4) 2.57 ☐ 2.05

5) 9.03 ☐ 0.930

6) 6.06 ☐ 6.6

7) 7.02 ☐ 7.020

8) 3.04 ☐ 3.2

9) 3.61 ☐ 3.245

10) 0.986 ☐ 0.0986

11) 17.24 ☐ 17.240

12) 0.759 ☐ 0.81

13) 9.040 ☐ 9.40

14) 5.73 ☐ 5.213

15) 9.44 ☐ 9.404

16) 7.17 ☐ 7.170

17) 4.85 ☐ 4.085

18) 9.041 ☐ 9.40

19) 3.033 ☐ 3.030

20) 4.97 ☐ 4.970

Rounding Decimals

✍ **Round each decimal to the nearest whole number.**

1) 28.12	3) 16.22	5) 7.95
2) 6.9	4) 8.5	6) 52.7

✍ **Round each decimal to the nearest tenth.**

7) 31.761	9) 94.729	11) 13.219
8) 14.421	10) 77.89	12) 59.89

✍ **Round each decimal to the nearest hundredth.**

13) 8.428	15) 55.3786	17) 62.241
14) 23.812	16) 231.912	18) 19.447

✍ **Round each decimal to the nearest thousandth.**

19) 15.54324	21) 243.8652	23) 67.1983
20) 34.62586	22) 80.4529	24) 72.36788

Answers of Worksheets

Simplifying Fractions

1) $\frac{1}{2}$ 7) $\frac{4}{5}$ 13) $\frac{7}{8}$ 19) $\frac{6}{31}$

2) $\frac{4}{5}$ 8) $\frac{1}{4}$ 14) $\frac{9}{10}$ 20) $\frac{4}{9}$

3) $\frac{3}{4}$ 9) $\frac{1}{2}$ 15) $\frac{1}{3}$ 21) $\frac{1}{8}$

4) $\frac{1}{2}$ 10) $\frac{5}{7}$ 16) $\frac{5}{14}$ 22) C

5) $\frac{1}{4}$ 11) $\frac{1}{4}$ 17) $\frac{2}{7}$ 23) A

6) $\frac{2}{3}$ 12) $\frac{1}{2}$ 18) $\frac{4}{27}$ 24) B

Adding and Subtracting Fractions

1) $\frac{9}{9} = 1$ 9) $1\frac{8}{21}$ 17) $-\frac{1}{24}$ 25) $\frac{3}{28}$

2) $\frac{9}{14}$ 10) $1\frac{1}{3}$ 18) $\frac{3}{28}$ 26) $\frac{27}{40}$

3) $\frac{5}{8}$ 11) $1\frac{7}{30}$ 19) $\frac{13}{18}$ 27) $\frac{18}{35}$

4) $1\frac{1}{10}$ 12) $\frac{3}{4}$ 20) $\frac{7}{12}$ 28) $\frac{5}{16}$

5) $\frac{17}{20}$ 13) $\frac{1}{6}$ 21) $\frac{19}{24}$ 29) $\frac{1}{2}$

6) $1\frac{1}{4}$ 14) $\frac{5}{8}$ 22) $-\frac{1}{15}$ 30) $\frac{1}{18}$

7) $1\frac{1}{5}$ 15) $\frac{1}{6}$ 23) $\frac{5}{28}$

8) $1\frac{1}{15}$ 16) $\frac{1}{20}$ 24) $\frac{1}{2}$

Multiplying and Dividing Fractions

1) $\frac{3}{5}$ 5) $\frac{1}{20}$ 9) $\frac{3}{8}$ 13) 2

2) $\frac{1}{3}$ 6) $\frac{1}{9}$ 10) $\frac{1}{14}$ 14) $\frac{1}{2}$

3) $\frac{1}{20}$ 7) $\frac{2}{7}$ 11) $\frac{1}{3}$ 15) $3\frac{3}{4}$

4) $\frac{1}{30}$ 8) $\frac{1}{16}$ 12) $\frac{1}{6}$ 16) $\frac{2}{5}$

17) $\frac{3}{4}$ 21) $\frac{7}{10}$ 25) $\frac{5}{18}$ 29) $3\frac{3}{4}$

18) $4\frac{1}{2}$ 22) 1 26) $\frac{25}{36}$ 30) $\frac{20}{33}$

19) $\frac{26}{49}$ 23) $\frac{3}{5}$ 27) $\frac{9}{10}$

20) $\frac{2}{9}$ 24) $\frac{3}{4}$ 28) $\frac{8}{15}$

Adding and Subtracting Mixed Numbers

1) $5\frac{1}{2}$ 8) $7\frac{9}{10}$ 15) $3\frac{1}{4}$ 22) $2\frac{1}{16}$

2) 8 9) $11\frac{29}{35}$ 16) $3\frac{13}{15}$ 23) $4\frac{1}{24}$

3) $4\frac{1}{2}$ 10) $14\frac{19}{48}$ 17) $3\frac{1}{8}$ 24) $2\frac{71}{72}$

4) $4\frac{7}{12}$ 11) $1\frac{1}{2}$ 18) $2\frac{13}{15}$ 25) $5\frac{33}{35}$

5) $6\frac{5}{12}$ 12) $2\frac{1}{5}$ 19) $3\frac{7}{30}$ 26) $6\frac{32}{63}$

6) $8\frac{13}{15}$ 13) $1\frac{2}{9}$ 20) $4\frac{3}{35}$

7) $6\frac{16}{21}$ 14) $1\frac{9}{14}$ 21) $5\frac{1}{12}$

Multiplying and Dividing Mixed Numbers

1) $12\frac{3}{8}$ 10) 11 19) $4\frac{4}{27}$

2) $23\frac{1}{9}$ 11) $\frac{4}{7}$ 20) $2\frac{26}{33}$

3) $35\frac{15}{16}$ 12) $1\frac{1}{4}$ 21) $3\frac{17}{36}$

4) $8\frac{2}{3}$ 13) $4\frac{2}{9}$ 22) $\frac{19}{22}$

5) 5 14) $2\frac{23}{24}$ 23) $3\frac{27}{91}$

6) $6\frac{6}{7}$ 15) 3 24) $3\frac{13}{36}$

7) $30\frac{1}{3}$ 16) $\frac{11}{45}$ 25) $7\frac{1}{5}$

8) $7\frac{6}{7}$ 17) $\frac{19}{84}$ 26) $4\frac{16}{45}$

9) $21\frac{7}{8}$ 18) $\frac{52}{57}$

Adding and Subtracting Decimals

1) 10.91   2) 76.87   3) 94.95   4) 17.04

5) 83.93	9) 25.52	13) 3.85	17) 46.58
6) 27.89	10) 2.6	14) 6.6	18) 9.69
7) 50.95	11) 1.78	15) 26.39	19) 52.8
8) 87.17	12) 1.55	16) 6.8	

Multiplying and Dividing Decimals

1) 0.3	7) 7.35	13) 0.185	19) 0.02215
2) 1.32	8) 85.1	14) 0.746	20) 76.5
3) 0.64	9) 104.72	15) 1.2	21) 1,611
4) 0.21	10) 50.7	16) 24	22) 281.5
5) 1.11	11) 48.3	17) 31	23) 96.75
6) 0.12	12) 352.8	18) 164	24) 931

Comparing Decimals

1) >	6) <	11) =	16) =
2) <	7) =	12) <	17) >
3) =	8) <	13) <	18) <
4) >	9) >	14) >	19) >
5) >	10) >	15) >	20) =

Rounding Decimals

1) 28	9) 94.7	17) 62.24
2) 7	10) 77.9	18) 19.45
3) 16	11) 13.2	19) 15.543
4) 9	12) 59.9	20) 34.626
5) 8	13) 8.43	21) 243.865
6) 53	14) 23.81	22) 80.453
7) 31.8	15) 55.38	23) 67.198
8) 14.4	16) 231.91	24) 72.368

Chapter 3 :

Proportions, Ratios, and Percent

Topics that you'll practice in this chapter:

- ✓ Simplifying Ratios
- ✓ Proportional Ratios
- ✓ Similarity and Ratios
- ✓ Ratio and Rates Word Problems
- ✓ Percentage Calculations
- ✓ Percent Problems
- ✓ Discount, Tax and Tip
- ✓ Percent of Change
- ✓ Simple Interest

Without mathematics, there's nothing you can do. Everything around you is mathematics.
Everything around you is numbers." – Shakuntala Devi

Simplifying Ratios

✍ **Reduce each ratio.**

1) $15 : 20 =$ ___ : ___   9) $8 : 48 =$ ___ : ___   17) $56 : 72 =$ ___ : ___

2) $7 : 70 =$ ___ : ___   10) $49 : 63 =$ ___ : ___   18) $26 : 13 =$ ___ : ___

3) $16 : 28 =$ ___ : ___   11) $18 : 27 =$ ___ : ___   19) $15 : 45 =$ ___ : ___

4) $7 : 21 =$ ___ : ___   12) $35 : 10 =$ ___ : ___   20) $28 : 4 =$ ___ : ___

5) $4 : 40 =$ ___ : ___   13) $90 : 9 =$ ___ : ___   21) $24 : 48 =$ ___ : ___

6) $6 : 48 =$ ___ : ___   14) $24 : 32 =$ ___ : ___   22) $30 : 24 =$ ___ : ___

7) $16 : 64 =$ ___ : ___   15) $7 : 56 =$ ___ : ___   23) $70 : 140 =$ ___ : ___

8) $10 : 25 =$ ___ : ___   16) $45 : 63 =$ ___ : ___   24) $6 : 180 =$ ___ : ___

✍ **Write each ratio as a fraction in simplest form.**

25) $6 : 12 =$ 32) $7 : 35 =$ 39) $20 : 300 =$

26) $30 : 50 =$ 33) $40 : 96 =$ 40) $30 : 120 =$

27) $15 : 35 =$ 34) $12 : 54 =$ 41) $56 : 42 =$

28) $9 : 27 =$ 35) $44 : 52 =$ 42) $26 : 130 =$

29) $8 : 24 =$ 36) $12 : 27 =$ 43) $66 : 123 =$

30) $18 : 84 =$ 37) $15 : 180 =$ 44) $70 : 630 =$

31) $7 : 14 =$ 38) $39 : 143 =$ 45) $75 : 125 =$

Proportional Ratios

✍ **Fill in the blanks; Calculate each proportion.**

1) $3:8 = $ __ $: 48$

2) $2:5 = 20:$ __

3) $1:9 = $ __ $: 81$

4) $6:7 = 12:$ __

5) $9:2 = 63:$ __

6) $8:7 = $ __ $: 49$

7) $20:3 = $ __ $: 15$

8) $1:3 = $ __ $: 75$

9) $7:6 = $ __ $: 60$

10) $8:5 = $ __ $: 45$

11) $3:10 = 60:$ __

12) $6:11 = 42:$ __

✍ **State if each pair of ratios form a proportion.**

13) $\frac{3}{20}$ and $\frac{9}{60}$

14) $\frac{1}{7}$ and $\frac{6}{42}$

15) $\frac{3}{7}$ and $\frac{24}{56}$

16) $\frac{4}{9}$ and $\frac{12}{18}$

17) $\frac{1}{9}$ and $\frac{12}{81}$

18) $\frac{7}{8}$ and $\frac{21}{28}$

19) $\frac{9}{13}$ and $\frac{27}{39}$

20) $\frac{1}{8}$ and $\frac{8}{64}$

21) $\frac{6}{19}$ and $\frac{30}{85}$

22) $\frac{5}{9}$ and $\frac{40}{81}$

23) $\frac{9}{14}$ and $\frac{108}{168}$

24) $\frac{15}{23}$ and $\frac{360}{552}$

✍ **Calculate each proportion.**

25) $\frac{20}{25} = \frac{32}{x}, x = $ ____

26) $\frac{1}{8} = \frac{32}{x}, x = $ ____

27) $\frac{15}{5} = \frac{21}{x}, x = $ ____

28) $\frac{1}{7} = \frac{x}{294}, x = $ ____

29) $\frac{7}{9} = \frac{x}{81}, x = $ ____

30) $\frac{1}{5} = \frac{13}{x}, x = $ ____

31) $\frac{9}{5} = \frac{36}{x}, x = $ ____

32) $\frac{6}{13} = \frac{48}{x}, x = $ ____

33) $\frac{5}{8} = \frac{x}{88}, x = $ ____

34) $\frac{4}{15} = \frac{x}{240}, x = $ ____

35) $\frac{9}{19} = \frac{x}{266}, x = $ ____

36) $\frac{7}{15} = \frac{x}{270}, x = $ ____

Similarity and Ratios

✍ **Each pair of figures is similar. Find the missing side.**

1)

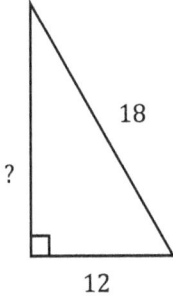

18

5

?

12

6

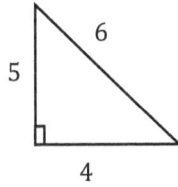

4

2)

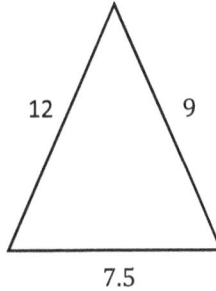

12 / 9

7.5

8 / 6

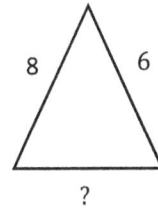

?

3)

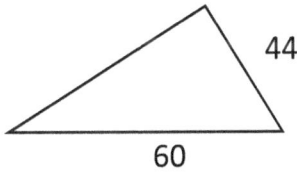

44

60

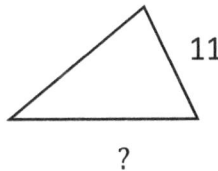

11

?

4)

104

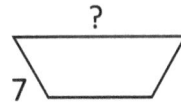

?

7

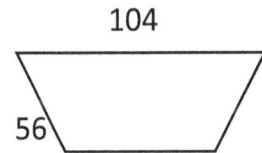

56

✍ **Calculate.**

5) Two rectangles are similar. The first is 24 feet wide and 120 feet long. The second is 30 feet wide. What is the length of the second rectangle?

_____

6) Two rectangles are similar. One is 5 meters by 36 meters. The longer side of the second rectangle is 90 meters. What is the other side of the second rectangle? _____

7) A building casts a shadow 25 ft long. At the same time a girl 10 ft tall casts a shadow 5 ft long. How tall is the building? _____

8) The scale of a map of Texas is 4 inches: 32 miles. If you measure the distance from Dallas to Martin County as 38.4 inches, approximately how far is Martin County from Dallas? _____

Ratio and Rates Word Problems

✍ **Find the answer for each word problem.**

1) Mason has 24 red cards and 36 green cards. What is the ratio of Mason 's red cards to his green cards? _____

2) In a party, 45 soft drinks are required for every 54 guests. If there are 378 guests, how many soft drinks is required? _____

3) In Mason's class, 42 of the students are tall and 24 are short. In Michael's class 84 students are tall and 48 students are short. Which class has a higher ratio of tall to short students? _____

4) The price of 5 apples at the Quick Market is $4.6. The price of 7 of the same apples at Walmart is $5.95. Which place is the better buy?

_____

5) The bakers at a Bakery can make 90 bagels in 3 hours. How many bagels can they bake in 24 hours? What is that rate per hour? _____

6) You can buy 5 cans of green beans at a supermarket for $5.75. How much does it cost to buy 45 cans of green beans? _____

7) The ratio of boys to girls in a class is 4: 7. If there are 32 boys in the class, how many girls are in that class? _____

8) The ratio of red marbles to blue marbles in a bag is 3: 7. If there are 50 marbles in the bag, how many of the marbles are red? _____

Percentage Calculations

✍ **Calculate the given percent of each value.**

1) 3% of 60 = ___

2) 20% of 32 = ___

3) 4% of 72 = ___

4) 16% of 32 = ___

5) 25% of 124 = ___

6) 35% of 56 = ___

7) 15% of 20 = ___

8) 14% of 150 = ___

9) 80% of 50 = ___

10) 12% of 115 = ___

11) 72% of 250 = ___

12) 52% of 500 = ___

13) 70% of 400 = ___

14) 27% of 145 = ___

15) 90% of 64 = ___

16) 60% of 55 = ___

17) 22% of 210 = ___

18) 8% of 235 = ___

✍ **Calculate the percent of each given value.**

19) ___% of 25 = 5

20) ___% of 40 = 20

21) ___% of 25 = 2

22) ___% of 50 = 16

23) ___% of 250 = 5

24) ___% of 40 = 32

25) ___% of 125 = 20

26) ___% of 700 = 49

27) ___% of 350 = 49

28) ___% of 500 = 210

✍ **Calculate each percent problem.**

29) A Cinema has 250 seats. 60 seats were sold for the current movie. What percent of seats are empty? _____ %

30) There are 68 boys and 92 girls in a class. 75% of the students in the class take the bus to school. How many students do not take the bus to school?

___

Percent Problems

✎ **Calculate each problem.**

1) 9 is what percent of 45? ___%

2) 60 is what percent of 120? ___%

3) 10 is what percent of 200? ___%

4) 15 is what percent of 125? ___%

5) 10 is what percent of 400? ___%

6) 66 is what percent of 55? ___%

7) 40 is what percent of 160? ___%

8) 40 is what percent of 50? ___%

9) 120 is what percent of 800? ___%

10) 78 is what percent of 120? ___%

11) 36 is what percent of 144? ___%

12) 17 is what percent of 85? ___%

13) 90 is what percent of 900? ___%

14) 36 is what percent of 16? ___%

15) 63 is what percent of 14? ___%

16) 18 is what percent of 60? ___%

17) 126 is what percent of 200? ___%

18) 232 is what percent of 40? ___%

✎ **Calculate each percent word problem.**

19) There are 40 employees in a company. On a certain day, 25 were present. What percent showed up for work? ___%

20) A metal bar weighs 60 ounces. 25% of the bar is gold. How many ounces of gold are in the bar? _____

21) A crew is made up of 12 women; the rest are men. If 15% of the crew are women, how many people are in the crew? _____

22) There are 40 students in a class and 8 of them are girls. What percent are boys? ___%

23) The Royals softball team played 400 games and won 280 of them. What percent of the games did they lose? ___%

Discount, Tax and Tip

✎ **Find the selling price of each item.**

1) Original price of a computer: $420

   Tax: 8% Selling price: $_____

2) Original price of a laptop: $280

   Tax: 4% Selling price: $_____

3) Original price of a sofa: $820

   Tax: 5% Selling price: $_____

4) Original price of a car: $15,800

   Tax: 3.6%   Selling price: $_____

5) Original price of a Table: $250

   Tax: 9% Selling price: $_____

6) Original price of a house: $630,000

   Tax: 1.8%   Selling price: $_____

7) Original price of a tablet: $450

   Discount: 30%   Selling price: $____

8) Original price of a chair: $390

   Discount: 8%   Selling price: $____

9) Original price of a book: $75

   Discount: 42%   Selling price: $____

10) Original price of a cellphone: $820

   Discount: 23%   Selling price: $___

11) Food bill: $45

   Tip: 15%   Price: $_____

12) Food bill: $32

   Tipp: 20%   Price: $_____

13) Food bill: $90

   Tip: 35%   Price: $_____

14) Food bill: $42

   Tipp: 12%   Price: $_____

✎ **Find the answer for each word problem.**

15) Nicolas hired a moving company. The company charged $500 for its services, and Nicolas gives the movers a 40% tip. How much does Nicolas tip the movers? $_____

16) Mason has lunch at a restaurant and the cost of his meal is $90. Mason wants to leave a 25% tip. What is Mason's total bill including tip? $_____

17) The sales tax in Texas is 19.80% and an item costs $350. How much is the tax? $_____

18) The price of a table at Best Buy is $680. If the sales tax is 5%, what is the final price of the table including tax? $_____

Percent of Change

✍ Find each percent of change.

1) From 150 to 450. ___ %

2) From 50 ft to 250 ft. ___ %

3) From $60 to $360. ___ %

4) From 60 cm to 180 cm. ___ %

5) From 15 to 45. ___ %

6) From 80 to 16. ___ %

7) From 120 to 360. ___ %

8) From 900 to 450. ___ %

9) From 1,000 to 200. ___ %

10) From 144 to 36. ___ %

✍ Calculate each percent of change word problem.

11) Bob got a raise, and his hourly wage increased from $42 to $63. What is the percent increase? ____ %

12) The price of a pair of shoes increases from $50 to $61. What is the percent increase? ___ %

13) At a coffee shop, the price of a cup of coffee increased from $4.80 to $5.76. What is the percent increase in the cost of the coffee? _____ %

14) 51 cm are cut from 85 cm board. What is the percent decrease in length? _____ %

15) In a class, the number of students has been increased from 54 to 81. What is the percent increase? _____ %

16) The price of gasoline rises from $24.40 to $30.50 in one month. By what percent did the gas price rise? _____ %

17) A shirt was originally priced at $38. It went on sale for $24.70. What was the percent that the shirt was discounted? _____ %

Simple Interest

✒ Determine the simple interest for these loans.

1) $480 at 11% for 3 years. $ _____

2) $4,200 at 7% for 4 years. $ _____

3) $2,500 at 20% for 3 years. $ _____

4) $6,800 at 3.9% for 4 months. $ ____

5) $800 at 6% for 7 months. $ _____

6) $36,000 at 4.2% for 6 years. $ _____

7) $6,500 at 7% for 4 years. $ _____

8) $850 at 9.5% for 2 years. $ _____

9) $1,200 at 5.8% for 9 months. $ ___

10) $3,000 at 4.5% for 7 years. $ _____

✒ Calculate each simple interest word problem.

11) A new car, valued at $22,000, depreciates at 8.5% per year. What is the value of the car one year after purchase? $_____

12) Sara puts $9,000 into an investment yielding 6% annual simple interest; she left the money in for three years. How much interest does Sara get at the end of those three years? $_____

13) A bank is offering 12% simple interest on a savings account. If you deposit $16,400, how much interest will you earn in two years? $_____

14) $720 interest is earned on a principal of $6,000 at a simple interest rate of 4% interest per year. For how many years was the principal invested? _____

15) In how many years will $2,200 yield an interest of $440 at 4% simple interest? _____

16) Jim invested $8,000 in a bond at a yearly rate of 4.5%. He earned $1,440 in interest. How long was the money invested? _____

Answers of Worksheets

Simplifying Ratios

1) $3:4$	14) $3:4$	26) $\frac{3}{5}$	36) $\frac{4}{9}$
2) $1:10$	15) $1:8$	27) $\frac{3}{7}$	37) $\frac{1}{12}$
3) $4:7$	16) $5:7$	28) $\frac{1}{3}$	38) $\frac{3}{11}$
4) $1:3$	17) $7:9$	29) $\frac{1}{3}$	39) $\frac{1}{15}$
5) $1:10$	18) $2:1$	30) $\frac{3}{14}$	40) $\frac{1}{4}$
6) $1:8$	19) $1:3$	31) $\frac{1}{2}$	41) $\frac{4}{3}$
7) $2:8$	20) $7:1$	32) $\frac{1}{5}$	42) $\frac{1}{5}$
8) $2:5$	21) $1:2$	33) $\frac{5}{12}$	43) $\frac{22}{41}$
9) $1:6$	22) $5:4$	34) $\frac{2}{9}$	44) $\frac{1}{9}$
10) $7:9$	23) $1:2$	35) $\frac{11}{13}$	45) $\frac{3}{5}$
11) $2:3$	24) $1:30$		
12) $7:2$	25) $\frac{1}{2}$		
13) $10:1$			

Proportional Ratios

1) 18	10) 72	19) Yes	28) 42
2) 50	11) 200	20) Yes	29) 63
3) 9	12) 77	21) No	30) 65
4) 14	13) Yes	22) No	31) 20
5) 14	14) Yes	23) Yes	32) 104
6) 56	15) Yes	24) Yes	33) 55
7) 100	16) No	25) 40	34) 64
8) 25	17) No	26) 256	35) 126
9) 70	18) No	27) 7	36) 126

Similarity and ratios

1) 15	4) 13	7) 50 feet
2) 5	5) 150 feet	8) 307.2 miles
3) 15	6) 12.5 meters	

Ratio and Rates Word Problems

1) $2:3$	2) 315

3) The ratio for both classes is 7 to 4.   6) $51.75

4) Walmart is a better buy.   7) 56

5) 720, the rate is 30 per hour.   8) 15

Percentage Calculations

1) 1.8	11) 180	21) 8%
2) 6.4	12) 260	22) 32%
3) 2.88	13) 280	23) 2%
4) 5.12	14) 39.15	24) 80%
5) 31	15) 57.6	25) 16%
6) 19.6	16) 33	26) 7%
7) 3	17) 46.2	27) 14%
8) 21	18) 18.8	28) 42%
9) 40	19) 20%	29) 76%
10) 13.8	20) 50%	30) 40

Percent Problems

1) 20%	9) 15%	17) 63%
2) 50%	10) 65%	18) 580%
3) 5%	11) 25%	19) 62.5%
4) 12%	12) 20%	20) 15 ounces
5) 2.5%	13) 10%	21) 80
6) 120%	14) 225%	22) 80%
7) 25%	15) 450%	23) 30%
8) 80%	16) 30%	

Discount, Tax and Tip

1) $453.60	7) $315.00	13) $121.50
2) $291.20	8) $358.80	14) $47.04
3) $861.00	9) $43.50	15) $200.00
4) $16,368.80	10) $631.40	16) $112.50
5) $272.50	11) $51.75	17) $69.30
6) $641,340	12) $38.40	18) $714.00

Percent of Change

1) 200%	7) 200%	13) 20%
2) 400%	8) 50%	14) 60%
3) 500%	9) 80%	15) 50%
4) 200%	10) 75%	16) 25%
5) 200%	11) 50%	17) 35%
6) 80%	12) 22%	

Simple Interest

1) $158.40	7) $1,820.00	13) $3,936.00
2) $1,176.00	8) $161.50	14) 3 years
3) $1,500.00	9) $52.20	15) 5 years
4) $88.40	10) $945.00	16) 4 years
5) $28.00	11) $20,130.00	
6) $9,072.00	12) $1,620.00	

Chapter 4 :

Exponents and Radicals Expressions

Topics that you'll practice in this chapter:

✓ Multiplication Property of Exponents

✓ Zero and Negative Exponents

✓ Division Property of Exponents

✓ Powers of Products and Quotients

✓ Negative Exponents and Negative Bases

✓ Scientific Notation

✓ Square Roots

✓ Simplifying Radical Expressions

✓ Simplifying Radical Expressions Involving Fractions

✓ Multiplying Radical Expressions

✓ Adding and Subtracting Radical Expressions

Love is anterior to life, posterior to death, initial of creation, and the exponent of breath.

Emily Dickinson

Multiplication Property of Exponents

✎ **Simplify and write the answer in exponential form.**

1) $4 \times 4^5 =$

2) $8^4 \times 8 =$

3) $7^3 \times 7^3 =$

4) $9^2 \times 9^2 =$

5) $2^2 \times 2^4 \times 2 =$

6) $5 \times 5^3 \times 5^3 =$

7) $4^3 \times 4^2 \times 4 \times 4 =$

8) $5x \times x =$

9) $x^3 \times x^3 =$

10) $x^7 \times x^2 =$

11) $x^4 \times x^3 \times x^2 =$

12) $10x \times 3x =$

13) $4x^3 \times 4x^3 =$

14) $7x^3 \times x =$

15) $3x^2 \times 4x^2 \times x^2 =$

16) $5x^4 \times x^4 =$

17) $2x^8 \times 2x =$

18) $6x \times x^5 =$

19) $4x^2 \times 6x^6 =$

20) $5yx^3 \times 4x =$

21) $7x^3 \times y^5x^7 =$

22) $y^2x^3 \times y^5x^4 =$

23) $3x^5 \times 4x^3y^4 =$

24) $4x^4 \times 9x^2y^5 =$

25) $5x^3y^4 \times 6x^8y^2 =$

26) $8x^3y^6 \times 4xy^3 =$

27) $2xy^5 \times 6x^3y^3 =$

28) $4x^5y^2 \times 4x^2y^8 =$

29) $7x \times 3y^8x^2 \times y^5 =$

30) $x^3 \times 2y^3x^4 \times 2y =$

31) $3yx^4 \times 3y^4x \times 3xy^3 =$

32) $6y^3 \times 2y^2x^4 \times 10yx^5 =$

Zero and Negative Exponents

✍ **Evaluate the following expressions.**

1) $1^{-5} =$

2) $4^{-1} =$

3) $0^{10} =$

4) $1^{15} =$

5) $5^{-2} =$

6) $3^{-3} =$

7) $9^{-1} =$

8) $10^{-2} =$

9) $12^{-2} =$

10) $2^{-5} =$

11) $3^{-4} =$

12) $2^{-4} =$

13) $6^{-3} =$

14) $10^{-3} =$

15) $30^{-1} =$

16) $15^{-2} =$

17) $4^{-3} =$

18) $2^{-7} =$

19) $5^{-3} =$

20) $4^{-4} =$

21) $3^{-5} =$

22) $10^{-4} =$

23) $2^{-10} =$

24) $8^{-3} =$

25) $20^{-2} =$

26) $14^{-2} =$

27) $9^{-3} =$

28) $100^{-2} =$

29) $5^{-4} =$

30) $4^{-6} =$

31) $\left(\frac{1}{4}\right)^{-3} =$

32) $\left(\frac{1}{6}\right)^{-2} =$

33) $\left(\frac{1}{7}\right)^{-2} =$

34) $\left(\frac{2}{3}\right)^{-3} =$

35) $\left(\frac{1}{13}\right)^{-2} =$

36) $\left(\frac{7}{12}\right)^{-2} =$

37) $\left(\frac{1}{6}\right)^{-3} =$

38) $\left(\frac{1}{300}\right)^{-2} =$

39) $\left(\frac{2}{9}\right)^{-2} =$

40) $\left(\frac{7}{5}\right)^{-1} =$

41) $\left(\frac{13}{23}\right)^{0} =$

42) $\left(\frac{1}{4}\right)^{-5} =$

Division Property of Exponents

✎ **Simplify.**

1) $\dfrac{5^6}{5^7} =$

2) $\dfrac{8^8}{8^6} =$

3) $\dfrac{4^5}{4} =$

4) $\dfrac{3}{3^5} =$

5) $\dfrac{x}{x^6} =$

6) $\dfrac{3 \times 3^2}{3^2 \times 3^5} =$

7) $\dfrac{9^4}{9^2} =$

8) $\dfrac{10 \times 10^9}{10^2 \times 10^7} =$

9) $\dfrac{7^5 \times 7^7}{7^4 \times 7^8} =$

10) $\dfrac{15x}{30x^6} =$

11) $\dfrac{3x^9}{4x^4} =$

12) $\dfrac{15x^8}{10x^9} =$

13) $\dfrac{42x^5}{6y^9} =$

14) $\dfrac{36y^8}{4x^4y^5} =$

15) $\dfrac{2x^7}{9x} =$

16) $\dfrac{49x^8y^6}{7x^9} =$

17) $\dfrac{48x^2}{24x^6y^{12}} =$

18) $\dfrac{30yx^5}{6yx^7} =$

19) $\dfrac{19x^7y}{38x^{12}y^4} =$

20) $\dfrac{9x^8}{63x^8} =$

21) $\dfrac{9x^{-9}}{4x^{-3}} =$

Powers of Products and Quotients

✍ **Simplify.**

1) $(4^3)^2 =$

2) $(2^3)^4 =$

3) $(2 \times 2^3)^2 =$

4) $(5 \times 5^5)^6 =$

5) $(19^4 \times 19^2)^3 =$

6) $(2^3 \times 2^4)^4 =$

7) $(5 \times 5^2)^2 =$

8) $(4^4)^4 =$

9) $(8x^5)^2 =$

10) $(3x^2y^4)^4 =$

11) $(7x^5y^2)^2 =$

12) $(5x^4y^4)^3 =$

13) $(2x^3y^3)^5 =$

14) $(10x^3y^4)^3 =$

15) $(13y^3y)^2 =$

16) $(5x^6x^4)^2 =$

17) $(6x^7y^6)^3 =$

18) $(12x^5x^7)^2 =$

19) $(2x^4 \times 2x)^4 =$

20) $(2x^4y^3)^5 =$

21) $(15x^7y^2)^2 =$

22) $(8x^3y^5)^3 =$

23) $(3x \times 2y^2)^4 =$

24) $\left(\frac{4x}{x^5}\right)^2 =$

25) $\left(\frac{x^4y^5}{x^3y^5}\right)^9 =$

26) $\left(\frac{36xy}{6x^5}\right)^3 =$

27) $\left(\frac{x^7}{x^8y^2}\right)^6 =$

28) $\left(\frac{xy^4}{x^3y^6}\right)^{-3} =$

29) $\left(\frac{5xy^8}{x^3}\right)^2 =$

30) $\left(\frac{xy^6}{2xy^3}\right)^{-4} =$

Negative Exponents and Negative Bases

✍ **Simplify.**

1) $-9^{-1} =$

2) $-9^{-2} =$

3) $-2^{-5} =$

4) $-x^{-7} =$

5) $11x^{-1} =$

6) $-8x^{-3} =$

7) $-12x^{-5} =$

8) $-9x^{-8}y^{-6} =$

9) $32x^{-5}y^{-1} =$

10) $10a^{-9}b^{-3} =$

11) $-17x^4y^{-6} =$

12) $-\dfrac{25}{x^{-5}} =$

13) $-\dfrac{13x}{a^{-7}} =$

14) $\left(-\dfrac{1}{3}\right)^{-4} =$

15) $\left(-\dfrac{3}{4}\right)^{-2} =$

16) $-\dfrac{14}{a^{-6}b^{-3}} =$

17) $-\dfrac{7x}{x^{-8}} =$

18) $-\dfrac{a^{-9}}{b^{-5}} =$

19) $-\dfrac{11}{x^{-5}} =$

20) $\dfrac{8b}{-16c^{-6}} =$

21) $\dfrac{12ab}{a^{-4}b^{-3}} =$

22) $-\dfrac{8n^{-4}}{32p^{-7}} =$

23) $\dfrac{16ab^{-6}}{-6c^{-5}} =$

24) $\left(\dfrac{10a}{5c}\right)^{-4} =$

25) $\left(-\dfrac{12x}{4yz}\right)^{-3} =$

26) $\dfrac{8ab^{-7}}{-5c^{-3}} =$

27) $\left(-\dfrac{x^4}{x^5}\right)^{-5} =$

28) $\left(-\dfrac{x^{-2}}{7x^3}\right)^{-2} =$

29) $\left(-\dfrac{x^{-4}}{x^2}\right)^{-6} =$

Scientific Notation

✎ **Write each number in scientific notation.**

1) 0.223 =

2) 0.09 =

3) 4.5 =

4) 900 =

5) 2,000 =

6) 0.006 =

7) 33 =

8) 9,400 =

9) 1,470 =

10) 52,000 =

11) 8,000,000 =

12) 0.00009 =

13) 2,158,000 =

14) 0.0039 =

15) 0.000075 =

16) 4,300,000 =

17) 130,000 =

18) 4,000,000,000 =

19) 0.00009 =

20) 0.0039 =

✎ **Write each number in standard notation.**

21) 4×10^{-1} =

22) 1.2×10^{-3} =

23) 2.7×10^5 =

24) 6×10^{-4} =

25) 3.6×10^{-3} =

26) 5.5×10^5 =

27) 3.2×10^4 =

28) 3.88×10^6 =

29) 7×10^{-6} =

30) 4.2×10^{-7} =

Square Roots

✎ **Find the value each square root.**

1) $\sqrt{16} =$ ___

2) $\sqrt{25} =$ ___

3) $\sqrt{1} =$ ___

4) $\sqrt{64} =$ ___

5) $\sqrt{0} =$ ___

6) $\sqrt{196} =$ ___

7) $\sqrt{4} =$ ___

8) $\sqrt{256} =$ ___

9) $\sqrt{36} =$ ___

10) $\sqrt{289} =$ ___

11) $\sqrt{169} =$ ___

12) $\sqrt{144} =$ ___

13) $\sqrt{100} =$ ___

14) $\sqrt{1,600} =$ ___

15) $\sqrt{2,500} =$ ___

16) $\sqrt{324} =$ ___

17) $\sqrt{529} =$ ___

18) $\sqrt{20} =$ ___

19) $\sqrt{625} =$ ___

20) $\sqrt{18} =$ ___

21) $\sqrt{50} =$ ___

22) $\sqrt{1,024} =$ ___

23) $\sqrt{160} =$ ___

24) $\sqrt{32} =$ ___

✎ **Evaluate.**

25) $\sqrt{4} \times \sqrt{25} =$ _____

26) $\sqrt{36} \times \sqrt{49} =$ _____

27) $\sqrt{6} \times \sqrt{6} =$ _____

28) $\sqrt{13} \times \sqrt{13} =$ _____

29) $2\sqrt{5} \times 3\sqrt{5} =$ _____

30) $\sqrt{12} \times \sqrt{3} =$ _____

31) $\sqrt{13} + \sqrt{13} =$ _____

32) $\sqrt{10} + 2\sqrt{10} =$ _____

33) $12\sqrt{7} - 10\sqrt{7} =$ _____

34) $4\sqrt{10} \times 2\sqrt{10} =$ _____

35) $5\sqrt{3} \times 8\sqrt{3} =$ _____

36) $6\sqrt{3} - \sqrt{12} =$ _____

Simplifying Radical Expressions

🖎 **Simplify.**

1) $\sqrt{13x^2} =$

2) $\sqrt{75x^2} =$

3) $\sqrt[3]{27a} =$

4) $\sqrt{64x^5} =$

5) $\sqrt{216a} =$

6) $\sqrt[3]{63w^3} =$

7) $\sqrt{192x} =$

8) $\sqrt{125v} =$

9) $\sqrt[3]{128x^2} =$

10) $\sqrt{100x^9} =$

11) $\sqrt{16x^4} =$

12) $\sqrt[3]{500a^5} =$

13) $\sqrt{242} =$

14) $\sqrt{392p^3} =$

15) $\sqrt{8m^6} =$

16) $\sqrt{198x^3y^3} =$

17) $\sqrt{121x^5y^5} =$

18) $\sqrt{16a^6b^3} =$

19) $\sqrt{90x^5y^7} =$

20) $\sqrt[3]{64y^2x^6} =$

21) $10\sqrt{16x^4} =$

22) $6\sqrt{81x^2} =$

23) $\sqrt[3]{56x^2y^6} =$

24) $\sqrt[3]{1,000x^5y^7} =$

25) $8\sqrt{50a} =$

26) $\sqrt[4]{625x^8y} =$

27) $\sqrt{24x^4y^5r^3} =$

28) $5\sqrt{36x^4y^5z^8} =$

29) $3\sqrt[3]{343x^9y^7} =$

30) $5\sqrt{81a^5b^2c^9} =$

31) $\sqrt[4]{625x^8y^{16}} =$

Multiplying Radical Expressions

✎ **Simplify.**

1) $\sqrt{5} \times \sqrt{5} =$

2) $\sqrt{5} \times \sqrt{10} =$

3) $\sqrt{3} \times \sqrt{12} =$

4) $\sqrt{49} \times \sqrt{47} =$

5) $\sqrt{7} \times -2\sqrt{28} =$

6) $3\sqrt{15} \times \sqrt{5} =$

7) $4\sqrt{72} \times \sqrt{2} =$

8) $\sqrt{5} \times -\sqrt{49} =$

9) $\sqrt{55} \times \sqrt{11} =$

10) $7\sqrt{42} \times 2\sqrt{216} =$

11) $\sqrt{45}(5 + \sqrt{5}) =$

12) $\sqrt{13x^2} \times \sqrt{13x^3} =$

13) $-2\sqrt{27} \times \sqrt{3} =$

14) $2\sqrt{13x^4} \times \sqrt{13x^4} =$

15) $\sqrt{14x^3} \times \sqrt{7x^2} =$

16) $-8\sqrt{5x} \times \sqrt{7x^5} =$

17) $-2\sqrt{16x^5} \times 4\sqrt{8x^3} =$

18) $-4\sqrt{32}(8 + \sqrt{32}) =$

19) $\sqrt{32x}(10 - \sqrt{2x}) =$

20) $\sqrt{2x}(8\sqrt{x^5} + \sqrt{8}) =$

21) $\sqrt{20r}(5 + \sqrt{5}) =$

22) $-4\sqrt{7x} \times 3\sqrt{14x^5} =$

23) $-2\sqrt{12x} \times 3\sqrt{2x}$

24) $-\sqrt{7v^3}(-3\sqrt{42v}) =$

25) $(\sqrt{11} - 5)(\sqrt{11} + 5) =$

26) $(-3\sqrt{5} + 3)(\sqrt{5} - 4) =$

27) $(4 - 6\sqrt{3})(-6 + \sqrt{3}) =$

28) $(8 - 3\sqrt{5})(7 - \sqrt{5}) =$

29) $(-1 - \sqrt{3x})(4 + \sqrt{3x}) =$

30) $(-5 + 2\sqrt{7r})(-5 + \sqrt{7r}) =$

31) $(-\sqrt{7n} + 1)(-\sqrt{7} - 5) =$

32) $(-3 + \sqrt{3})(5 - 2\sqrt{3x}) =$

Simplifying Radical Expressions Involving Fractions

✍ *Simplify.*

1) $\dfrac{\sqrt{5}}{\sqrt{3}} =$

2) $\dfrac{\sqrt{18}}{\sqrt{45}} =$

3) $\dfrac{\sqrt{10}}{5\sqrt{2}} =$

4) $\dfrac{13}{\sqrt{3}} =$

5) $\dfrac{12\sqrt{5r}}{\sqrt{m^5}} =$

6) $\dfrac{11\sqrt{2}}{\sqrt{k}} =$

7) $\dfrac{6\sqrt{20x^3}}{\sqrt{16x}} =$

8) $\dfrac{\sqrt{14x^3y^4}}{\sqrt{7x^4y^3}} =$

9) $\dfrac{1}{1-\sqrt{5}} =$

10) $\dfrac{1-8\sqrt{a}}{\sqrt{11a}} =$

11) $\dfrac{\sqrt{a}}{\sqrt{a}+\sqrt{b}} =$

12) $\dfrac{1-\sqrt{5}}{2-\sqrt{6}} =$

13) $\dfrac{4+\sqrt{7}}{3-\sqrt{8}} =$

14) $\dfrac{5}{-3-3\sqrt{3}} =$

15) $\dfrac{7}{2-\sqrt{5}} =$

16) $\dfrac{\sqrt{7}-\sqrt{3}}{\sqrt{3}-\sqrt{7}} =$

17) $\dfrac{\sqrt{5}+\sqrt{7}}{\sqrt{7}-\sqrt{5}} =$

18) $\dfrac{2\sqrt{2}-\sqrt{3}}{3\sqrt{2}+\sqrt{5}} =$

19) $\dfrac{\sqrt{11}+5\sqrt{3}}{4-\sqrt{11}} =$

20) $\dfrac{\sqrt{5}+\sqrt{3}}{2-\sqrt{3}} =$

21) $\dfrac{\sqrt{32a^7b^4}}{\sqrt{2ab^3}} =$

22) $\dfrac{10\sqrt{21x^5}}{5\sqrt{x^3}} =$

Adding and Subtracting Radical Expressions

✎ **Simplify.**

1) $\sqrt{2} + \sqrt{8} =$

2) $3\sqrt{50} + 4\sqrt{2} =$

3) $2\sqrt{12} - 4\sqrt{3} =$

4) $5\sqrt{32} - 5\sqrt{2} =$

5) $3\sqrt{75} - 5\sqrt{3} =$

6) $-\sqrt{72} - 4\sqrt{2} =$

7) $-7\sqrt{16} - 4\sqrt{25} =$

8) $8\sqrt{24} + 2\sqrt{6} =$

9) $10\sqrt{49} - 7\sqrt{100} =$

10) $-7\sqrt{5} + 9\sqrt{45} =$

11) $-15\sqrt{12} + 14\sqrt{48} =$

12) $20\sqrt{4} - 2\sqrt{25} =$

13) $-2\sqrt{20} + 7\sqrt{5} =$

14) $8\sqrt{7} - 2\sqrt{63} =$

15) $5\sqrt{44} + 3\sqrt{11} =$

16) $3\sqrt{27} - 5\sqrt{48} =$

17) $\sqrt{144} - \sqrt{81} =$

18) $3\sqrt{20} - 6\sqrt{5} =$

19) $-2\sqrt{7} + 8\sqrt{28} =$

20) $3\sqrt{75} - 2\sqrt{3} =$

21) $5\sqrt{27} - 3\sqrt{3} =$

22) $-7\sqrt{30} + 6\sqrt{120} =$

23) $-7\sqrt{24} - 2\sqrt{6} =$

24) $-\sqrt{32x} + 4\sqrt{2x} =$

25) $\sqrt{7y^2} + y\sqrt{112} =$

26) $\sqrt{45mn^2} + 2n\sqrt{5m} =$

27) $-4\sqrt{12a} - 4\sqrt{3a} =$

28) $-5\sqrt{15ab} - 2\sqrt{60ab} =$

29) $\sqrt{45x^2y} + x\sqrt{20y} =$

30) $2\sqrt{7a} + 4\sqrt{63a} =$

Answers of Worksheets

Multiplication Property of Exponents

1) 4^6
2) 8^5
3) 7^6
4) 9^4
5) 2^7
6) 5^7
7) 4^7
8) $5x^2$

9) x^6
10) x^9
11) x^9
12) $30x^2$
13) $16x^6$
14) $7x^4$
15) $12x^6$
16) $5x^8$

17) $4x^9$
18) $6x^6$
19) $24x^8$
20) $20x^4y$
21) $7x^{10}y^5$
22) x^7y^7
23) $12x^8y^4$
24) $36x^6y^5$

25) $30x^{11}y^6$
26) $32x^4y^9$
27) $12x^4y^8$
28) $16x^7y^{10}$
29) $21x^3y^{13}$
30) $4x^7y^4$
31) $27x^6y^8$
32) $120x^9y^6$

Zero and Negative Exponents

1) 1
2) $\frac{1}{4}$
3) 0
4) 1
5) $\frac{1}{25}$
6) $\frac{1}{27}$
7) $\frac{1}{9}$
8) $\frac{1}{100}$
9) $\frac{1}{144}$
10) $\frac{1}{32}$
11) $\frac{1}{81}$

12) $\frac{1}{16}$
13) $\frac{1}{216}$
14) $\frac{1}{1,000}$
15) $\frac{1}{30}$
16) $\frac{1}{225}$
17) $\frac{1}{64}$
18) $\frac{1}{128}$
19) $\frac{1}{125}$
20) $\frac{1}{256}$
21) $\frac{1}{243}$

22) $\frac{1}{10,000}$
23) $\frac{1}{1,024}$
24) $\frac{1}{512}$
25) $\frac{1}{400}$
26) $\frac{1}{196}$
27) $\frac{1}{729}$
28) $\frac{1}{10,000}$
29) $\frac{1}{625}$
30) $\frac{1}{4,096}$
31) 64
32) 36

33) 49
34) $\frac{27}{8}$
35) 169
36) $\frac{144}{49}$
37) 216
38) $90,000$
39) $\frac{81}{4}$
40) $\frac{5}{7}$
41) 1
42) $1,024$

Division Property of Exponents

1) $\frac{1}{5}$
2) 8^2
3) 4^4
4) $\frac{1}{3^4}$

5) $\frac{1}{x^5}$
6) $\frac{1}{3^4}$
7) 9^2
8) 10

9) 1
10) $\frac{1}{2x^5}$
11) $\frac{3x^5}{4}$
12) $\frac{3}{2x}$

13) $\frac{7x^5}{y^9}$
14) $\frac{9y^3}{x^4}$
15) $\frac{2x^6}{9}$

16) $\frac{7y^6}{x}$ 18) $\frac{5}{x^2}$ 19) $\frac{1}{2x^5y^3}$ 21) $\frac{9}{4x^6}$

17) $\frac{2}{x^4y^{12}}$ 20) $\frac{1}{7}$

Powers of Products and Quotients

1) 4^6

2) 2^{12}

3) 2^8

4) 5^{36}

5) 19^{18}

6) 2^{28}

7) 5^6

8) 4^{16}

9) $64x^{10}$

10) $81x^8y^{16}$

11) $49x^{10}y^4$

12) $125x^{12}y^{12}$

13) $32x^{15}y^{15}$

14) $1,000x^9y^{12}$

15) $169y^8$

16) $25x^{20}$

17) $216x^{21}y^{18}$

18) $144x^{24}$

19) $256x^{20}$

20) $32x^{20}y^{15}$

21) $225x^{14}y^4$

22) $512x^9y^{15}$

23) $1,296x^4y^8$

24) $\frac{16}{x^8}$

25) x^9

26) $\frac{216y^3}{x^{12}}$

27) $\frac{1}{x^6y^{12}}$

28) x^6y^6

29) $\frac{25y^{16}}{x^4}$

30) $\frac{16}{y^{12}}$

Negative Exponents and Negative Bases

1) $-\frac{1}{9}$

2) $-\frac{1}{81}$

3) $-\frac{1}{32}$

4) $-\frac{1}{x^7}$

5) $\frac{11}{x}$

6) $-\frac{8}{x^3}$

7) $-\frac{12}{x^5}$

8) $-\frac{9}{x^8y^6}$

9) $\frac{32}{x^5y}$

10) $\frac{10}{a^9b^3}$

11) $-\frac{17x^4}{y^6}$

12) $-25x^5$

13) $-13xa^7$

14) 81

15) $\frac{16}{9}$

16) $-14a^6b^3$

17) $-7x^9$

18) $-\frac{b^5}{a^9}$

19) $-11x^5$

20) $-\frac{bc^6}{2}$

21) $12a^5b^4$

22) $-\frac{p^7}{4n^4}$

23) $-\frac{8ac^5}{3b^6}$

24) $\frac{c^4}{16a^4}$

25) $\frac{y^3z^3}{27x^3}$

26) $-\frac{8ac^3}{5b^7}$

27) $-x^5$

28) $49x^{10}$

29) x^{36}

Scientific Notation

1) 2.23×10^{-1}
2) 9×10^{-2}
3) 4.5×10^0
4) 9×10^2
5) 2×10^3
6) 6×10^{-3}
7) 3.3×10^1
8) 9.4×10^3
9) 1.47×10^3
10) 5.2×10^4

11) 8×10^6
12) 9×10^{-5}
13) 2.158×10^6
14) 3.9×10^{-3}
15) 7.5×10^{-5}
16) 4.3×10^6
17) 1.3×10^5
18) 4×10^9
19) 9×10^{-5}
20) 3.9×10^{-3}

21) 0.4
22) 0.0012
23) 270,000
24) 0.0006
25) 0.0036
26) 550,000
27) 32,000
28) 3,880,000
29) 0.000007
30) 0.00000042

Square Roots

1) 4
2) 5
3) 1
4) 8
5) 0
6) 14
7) 2
8) 16
9) 6

10) 17
11) 13
12) 12
13) 10
14) 40
15) 50
16) 18
17) 23
18) $2\sqrt{5}$

19) 25
20) $3\sqrt{2}$
21) $5\sqrt{2}$
22) 32
23) $4\sqrt{10}$
24) $4\sqrt{2}$
25) 10
26) 42
27) 6

28) 13
29) 30
30) 6
31) $2\sqrt{13}$
32) $3\sqrt{10}$
33) $2\sqrt{7}$
34) 80
35) 120
36) $4\sqrt{3}$

Simplifying radical expressions

1) $x\sqrt{13}$
2) $5x\sqrt{3}$
3) $3\sqrt[3]{a}$
4) $8x^2\sqrt{x}$
5) $6\sqrt{6a}$
6) $w\sqrt[3]{63}$
7) $8\sqrt{3x}$
8) $5\sqrt{5v}$

9) $4\sqrt[3]{2x^2}$
10) $10x^4\sqrt{x}$
11) $4x^2$
12) $5a\sqrt[3]{4a^2}$
13) $11\sqrt{2}$
14) $14p\sqrt{2p}$
15) $2m^3\sqrt{2}$
16) $3x.y\sqrt{22xy}$

17) $11x^2y^2\sqrt{xy}$
18) $4a^3b\sqrt{b}$
19) $3x^2y^3\sqrt{10xy}$
20) $4x^2\sqrt[3]{y^2}$
21) $40x^2$
22) $54x$
23) $2y^2\sqrt[3]{7x^2}$
24) $10xy^2\sqrt[3]{x^2y}$

25) $40\sqrt{2a}$

26) $5x^2\sqrt[4]{y}$

27) $2x^2y^2\text{r}\sqrt{6yr}$

28) $30x^2y^2z^4\sqrt{y}$

29) $21x^3y^2\sqrt[3]{y}$

30) $45a^2bc^4\sqrt{ac}$

31) $5x^2y^4$

Multiplying radical expressions

1) 5

2) $5\sqrt{2}$

3) 6

4) $7\sqrt{47}$

5) -28

6) $15\sqrt{3}$

7) 48

8) $-5\sqrt{7}$

9) $11\sqrt{5}$

10) $504\sqrt{7}$

11) $15\sqrt{5} + 15$

12) $13x^2\sqrt{x}$

13) -18

14) $26x^4$

15) $7x^2\sqrt{2x}$

16) $-8x^3\sqrt{35}$

17) $-64x^4\sqrt{2}$

18) $-128\sqrt{2} - 128$

19) $40\sqrt{2x} - 8x$

20) $8x^3\sqrt{2} + 4\sqrt{x}$

21) $10\sqrt{5r} + 10\sqrt{r}$

22) $-84x^3\sqrt{2}$

23) $-12\sqrt{6}x$

24) $21v^2\sqrt{6}$

25) -14

26) $15\sqrt{5} - 27$

27) $40\sqrt{3} - 42$

28) $71 - 29\sqrt{5}$

29) $-3x - 5\sqrt{3x} - 4$

30) $14r - 15\sqrt{7r} + 25$

31) $7\sqrt{n} + 5\sqrt{7n} - \sqrt{7} - 5$

32) $-15 + 6\sqrt{3x} + 5\sqrt{3} - 6\sqrt{x}$

Simplifying radical expressions involving fractions

1) $\frac{\sqrt{15}}{3}$

2) $\frac{9\sqrt{10}}{45} = \frac{\sqrt{10}}{5}$

3) $\frac{\sqrt{20}}{10} = \frac{\sqrt{5}}{5}$

4) $\frac{13\sqrt{3}}{3}$

5) $\frac{12\sqrt{5mr}}{m^3}$

6) $\frac{11\sqrt{2k}}{k}$

7) $3x\sqrt{5}$

8) $\frac{\sqrt{2x}}{xy}$

9) $\frac{-1-\sqrt{5}}{4}$

10) $\frac{\sqrt{11a}-8a\sqrt{11}}{11a}$

11) $\frac{a-\sqrt{ab}}{a-b}$

12) $\frac{\sqrt{30}+2\sqrt{5}-\sqrt{6}-2}{2}$

13) $12+8\sqrt{2}+3\sqrt{7}+2\sqrt{14}$

14) $-\frac{5(\sqrt{3}-1)}{6}$

15) $-14-7\sqrt{5}$

16) -1

17) $6+\sqrt{35}$

18) $\frac{12-2\sqrt{10}-3\sqrt{6}+\sqrt{15}}{13}$

19) $\frac{4\sqrt{11}+11+20\sqrt{3}+5\sqrt{33}}{5}$

20) $2\sqrt{5}+3+\sqrt{15}+2\sqrt{3}$

21) $4a^3\sqrt{b}$

22) $2x\sqrt{21}$

Adding and subtracting radical expressions

1) $3\sqrt{2}$

2) $19\sqrt{2}$

3) 0

4) $15\sqrt{2}$

5) $10\sqrt{3}$

6) $-10\sqrt{2}$

7) -48

8) $18\sqrt{6}$

9) 0

10) $20\sqrt{5}$

11) $26\sqrt{3}$

12) 30

13) $3\sqrt{5}$

14) $2\sqrt{7}$

15) $13\sqrt{11}$

16) $-11\sqrt{3}$

17) 3

18) 0

19) $14\sqrt{7}$

20) $13\sqrt{3}$

21) $12\sqrt{3}$

22) $5\sqrt{30}$

23) $-16\sqrt{6}$

24) 0

25) $5y\sqrt{7}$

26) $5n\sqrt{5m}$

27) $-12\sqrt{3a}$

28) $-9\sqrt{15ab}$

29) $5x\sqrt{5y}$

30) $14\sqrt{7a}$

Chapter 5 :

Algebraic Expressions

Topics that you'll practice in this chapter:

- ✓ Simplifying Variable Expressions
- ✓ Simplifying Polynomial Expressions
- ✓ Translate Phrases into an Algebraic Statement
- ✓ The Distributive Property
- ✓ Evaluating One Variable Expressions
- ✓ Evaluating Two Variables Expressions
- ✓ Combining like Terms

I want freedom for the full expression of my personality.

Mahatma Gandhi

Simplifying Variable Expressions

✎ **Simplify each expression.**

1) $3(x + 5) =$

2) $(-4)(7x - 5) =$

3) $11x + 5 - 6x =$

4) $-4 - 2x^2 - 6x^2 =$

5) $7 + 13x^2 + 3 =$

6) $3x^2 + 7x + 15x^2 =$

7) $3x^2 - 12x^2 + 4x =$

8) $4x^2 - 8x - 2x =$

9) $6x + 7(3 - 4x) =$

10) $8x + 4(15x - 3) =$

11) $6(-3x - 9) - 17 =$

12) $-11x^2 - (-5x) =$

13) $2x + 7 + 5 - 8x =$

14) $7 + 6x - 11 - 5x =$

15) $27x + 8 - 13 - 5x =$

16) $(-11)(-5x + 2) - 41x =$

17) $19x - 4(4 - 2x) =$

18) $16x + 3(3x + 6) + 10 =$

19) $5(-2x - 4) - 13x =$

20) $16x - 3x(x + 10) =$

21) $17x + 5x(2 - 4x) =$

22) $5x(-4x - 7) + 20x =$

23) $25x - 19 + 4x^2 =$

24) $6x(x - 11) + 25 =$

25) $4x - 5 + 15x + 3x^2 =$

26) $-7x^2 - 11x - 9x =$

27) $10x - 9x^2 - 3x^2 - 7 =$

28) $13 + 3x^2 - 9x^2 - 21x =$

29) $22x + 10x^2 - 15x + 17 =$

30) $4x^2 + 25x + 21x^2 =$

31) $29 - 12x^2 - 23x - 4x^2 =$

32) $22x - 19x - 9x^2 + 30 =$

Simplifying Polynomial Expressions

✎ **Simplify each polynomial.**

1) $(2x^3 + 8x^2) - (11x + 3x^2) =$ _____

2) $(2x^5 + 7x^3) - (5x^3 + 11x^2) =$ _____

3) $(41x^4 + 5x^2) - (4x^2 + 20x^4) =$ _____

4) $13x - 8x^2 + 4(4x^2 + 3x^3) =$ _____

5) $(4x^3 - 22) + 5(3x^2 - 6x^3) =$ _____

6) $(4x^3 - 3x) - 5(2x^3 + x^4) =$ _____

7) $5(5x - 2x^3) - 2(8x^3 + 5x^2) =$ _____

8) $(3x^2 - 10x) - (5x^3 + 14x^2) =$ _____

9) $5x^3 - (3x^4 + 5x) + 2x^2 =$ _____

10) $11x^4 - (3x^2 + 5x) + 7x =$ _____

11) $(6x^2 - 3x^4) - (10x^4 + 3x^2) =$ _____

12) $2x^2 - 7x^3 + 19x^4 - 22x^3 =$ _____

13) $10x^2 - x^4 + 4x^4 - 32x^3 =$ _____

14) $-5x^2 + 17x^3 - 8x^2 - 6x =$ _____

15) $x^4 - 11x^5 - 30x^4 + 5x^2 =$ _____

16) $21x^3 + 13x - 5x^2 - 11x^3 =$ _____

Translate Phrases into an Algebraic Statement

✍ **Write an algebraic expression for each phrase.**

1) 9 multiplied by x. _____

2) Subtract 11 from y. _____

3) 19 divided by x. _____

4) 38 decreased by y. _____

5) Add y to 40. _____

6) The square of 6. _____

7) x raised to the fifth power. _____

8) The sum of six and a number. _____

9) The difference between fifty–seven and y. _____

10) The quotient of nine and a number. _____

11) The quotient of the square of x and 25. _____

12) The difference between x and 6 is 19. _____

13) 10 times a reduced by the square of b. _____

14) Subtract the product of a and b from 41. _____

The Distributive Property

✍ **Use the distributive property to simply each expression.**

1) $4(1 + 2x) =$

2) $2(4 + 7x) =$

3) $3(4x - 4) =$

4) $(2x - 5)(-6) =$

5) $(-3)(x + 6) =$

6) $(4 + 3x)2 =$

7) $(-5)(8 - 3x) =$

8) $-(-5 - 7x) =$

9) $(-6x + 3)(-3) =$

10) $(-4)(x - 7) =$

11) $-(5 - 3x) =$

12) $3(9 + 4x) =$

13) $6(4 + 3x) =$

14) $(-5x + 3)2 =$

15) $(5 - 8x)(-3) =$

16) $(-12)(3x + 3) =$

17) $(5 - 3x)6 =$

18) $4(2 + 6x) =$

19) $8(7x - 3) =$

20) $(-2x + 3)4 =$

21) $(7 - 5x)(-9) =$

22) $(-10)(x - 8) =$

23) $(11 - 4x)3 =$

24) $(-6)(10x - 4) =$

25) $(3 - 9x)(-7) =$

26) $(-9)(x + 9) =$

27) $(-3 + 5x)(-7) =$

28) $(-5)(8 - 10x) =$

29) $12(4x - 8) =$

30) $(-10x + 13)(-3) =$

31) $(-8)(3x - 2) + 4(x + 5) =$

32) $(-8)(x + 4) - (6 + 5x) =$

Evaluating One Variable Expressions

✍ **Evaluate each expression using the value given.**

1) $8 - x$, $x = 5$

2) $x - 9$, $x = 5$

3) $5x + 4$, $x = 3$

4) $x - 13$, $x = -4$

5) $12 - x$, $x = 4$

6) $x + 2$, $x = 6$

7) $4x + 8$, $x = 3$

8) $x + (-7)$, $x = -8$

9) $4x + 5$, $x = 2$

10) $3x + 9$, $x = -2$

11) $15 + 3x - 7$, $x = 2$

12) $17 - 3x$, $x = 3$

13) $8x - 9$, $x = 4$

14) $5x + 4$, $x = -3$

15) $10x + 5$, $x = 3$

16) $14 - 4x$, $x = -6$

17) $3(5x + 3)$, $x = 9$

18) $4(-3x - 6)$, $x = 3$

19) $7x - 2x + 12$, $x = 4$

20) $(5x + 6) \div 2$, $x = 8$

21) $(x + 18) \div 10$, $x = 12$

22) $5x - 12 + 3x$, $x = -3$

23) $(6 - 4x)(-3)$, $x = -4$

24) $9x^2 + 3x - 6$, $x = 2$

25) $x^2 - 10x$, $x = -5$

26) $3x(7 - 2x)$, $x = 2$

27) $12x + 6 - 2x^2$, $x = -4$

28) $(-3)(4x - 8 + 3x)$, $x = 3$

29) $(-6) + \frac{x}{4} + 3x$, $x = 16$

30) $(-6) + \frac{x}{5}$, $x = 35$

31) $\left(-\frac{45}{x}\right) - 7 + 2x$, $x = 9$

32) $\left(-\frac{21}{x}\right) - 12 + 4x$, $x = 7$

Evaluating Two Variables Expressions

✎ Evaluate each expression using the values given.

1) $2x - 4y$,

   $x = 4, y = 1$

2) $3x + 5y$,

   $x = -2, y = 2$

3) $-7a + 4b$,

   $a = 2, b = 4$

4) $3x + 5 - y$,

   $x = 5, y = 6$

5) $3z + 12 - 2k$,

   $z = 5, k = 6$

6) $6(-x - 3y)$,

   $x = 5, y = -2$

7) $5a + 3b$,

   $a = 3, b = 4$

8) $7x \div 3y$,

   $x = 3, y = 7$

9) $2x + 15 + 5y$,

   $x = -3, y = 1$

10) $5a - (18 - b)$,

   $a = 2, b = 8$

11) $2z + 20 + 5k$,

   $z = -6, k = 5$

12) $xy + 10 + 4x$,

   $x = 3, y = 5$

13) $2x + 4y - 8 + 5$,

   $x = 5, y = 2$

14) $\left(-\frac{24}{x}\right) + 3 + 2y$,

   $x = 4, y = 6$

15) $(-3)(-3a - 3b)$,

   $a = 4, b = 5$

16) $12 + 4x - 7 - y$,

   $x = 3, y = 5$

17) $11x + 5 - 8y + 6$,

   $x = 5, y = 2$

18) $10 + 2(-4x - 5y)$,

   $x = 5, y = 4$

19) $5x + 13 + 6y$,

   $x = 5, y = 6$

20) $10a - (7a + 3b) - 11$,

   $a = 3, b = 8$

Combining like Terms

✎ **Simplify each expression.**

1) $11x + 3x + 6 =$

2) $8(2x - 6) =$

3) $18x - 7x + 11 =$

4) $(-4)(6x - 7) =$

5) $22x - 10x - 5 =$

6) $32x - 13 + 8x =$

7) $15 - (8x - 11) =$

8) $-24x + 17 - 11x =$

9) $12x - 8 - 6x + 9 =$

10) $21x + 5 - 36 + 12x =$

11) $28x + 3x - 11 =$

12) $(-3x + 4)5 =$

13) $2 + 4x + 9x - 8 =$

14) $6(2x - 5x) - 4 =$

15) $4(5x + 11) + 3x =$

16) $x - 14 - 11x =$

17) $5(10 + 9x) - 8x =$

18) $42x + 17 - 23x =$

19) $(-7x) + 19 + 20x =$

20) $(-7x) - 33 + 29x =$

21) $4(5x + 3) - 19x =$

22) $5(6 - 2x) - 15x =$

23) $-24x + (11 - 18x) =$

24) $(-9) - (6)(7x + 3) =$

25) $(-1)(8x - 10) - 21x =$

26) $-36x + 14 + 27x - 5x =$

27) $3(-13x + 6) - 17x =$

28) $-5x - 42 + 32x =$

29) $37x - 19x + 15 - 9x =$

30) $3(5x + 7x) - 31 =$

31) $14 - 6x - 15 - 9x =$

32) $-2(-5x - 7x) + 27x =$

Answers of Worksheets

Simplifying Variable Expressions

1) $3x + 15$

2) $-28x + 20$

3) $5x + 5$

4) $-8x^2 - 4$

5) $13x^2 + 10$

6) $18x^2 + 7x$

7) $-9x^2 + 4x$

8) $4x^2 - 10x$

9) $-22x + 21$

10) $68x - 12$

11) $-18x - 71$

12) $-11x^2 + 5x$

13) $-6x + 12$

14) $x - 4$

15) $22x - 5$

16) $14x - 22$

17) $27x - 16$

18) $25x + 28$

19) $-23x - 20$

20) $-3x^2 - 14x$

21) $-20x^2 + 27x$

22) $-20x^2 - 15x$

23) $4x^2 + 25x - 19$

24) $6x^2 - 66x + 25$

25) $3x^2 + 19x - 5$

26) $-7x^2 - 20x$

27) $-12x^2 + 10x - 7$

28) $-6x^2 - 21x + 13$

29) $10x^2 + 7x + 17$

30) $25x^2 + 25x$

31) $-16x^2 - 23x + 29$

32) $-9x^2 + 3x + 30$

Simplifying Polynomial Expressions

1) $2x^3 + 5x^2 - 11x$

2) $2x^5 + 2x^3 - 11x^2$

3) $21x^4 + x^2$

4) $12x^3 + 8x^2 + 13x$

5) $-26x^3 + 15x^2 - 22$

6) $-5x^4 - 6x^3 - 3x$

7) $-26x^3 - 10x^2 + 25x$

8) $-5x^3 - 11x^2 - 10x$

9) $-3x^4 + 5x^3 + 2x^2 - 5x$

10) $11x^4 - 3x^2 + 2x$

11) $-13x^4 + 3x^2$

12) $19x^4 - 29x^3 + 2x^2$

13) $3x^4 - 32x^3 + 10x^2$

14) $17x^3 - 13x^2 - 6x$

15) $-11x^5 - 29x^4 + 5x^2$

16) $10x^3 - 5x^2 + 13x$

Translate Phrases into an Algebraic Statement

1) $9x$

2) $y - 11$

3) $\frac{19}{x}$

4) $38 - y$

5) $y + 40$

6) 6^2

7) x^5

8) $6 + x$

9) $57 - y$

10) $\frac{9}{x}$

11) $\frac{x^2}{25}$

12) $x - 6 = 19$

13) $10a - b^2$

14) $41 - ab$

The Distributive Property

1) $8x + 4$

2) $14x + 8$

3) $12x - 12$

4) $-12x + 30$

5) $-3x - 18$

6) $6x + 8$

7) $15x - 40$

8) $7x + 5$

9) $18x - 9$

10) $-4x + 28$

11) $3x - 5$

12) $12x + 27$

13) $18x + 24$	18) $24x + 8$	23) $-12x + 33$	28) $50x - 40$
14) $-10x + 6$	19) $56x - 24$	24) $-60x + 24$	29) $48x - 96$
15) $24x - 15$	20) $-8x + 12$	25) $63x - 21$	30) $30x - 39$
16) $-36x - 36$	21) $45x - 63$	26) $-9x - 81$	31) $-20x + 36$
17) $-18x + 30$	22) $-10x + 80$	27) $-35x + 21$	32) $-13x - 38$

Evaluating One Variables

1) 3	9) 13	17) 144	25) 75
2) -4	10) 3	18) -60	26) 18
3) 19	11) 14	19) 32	27) -74
4) -17	12) 8	20) 23	28) -39
5) 8	13) 23	21) 3	29) 46
6) 8	14) -11	22) -36	30) 1
7) 20	15) 35	23) -66	31) 6
8) -15	16) 38	24) 36	32) 13

Evaluating Two Variables

1) 4	6) 6	11) 33	16) 12
2) 4	7) 27	12) 37	17) 50
3) 2	8) 1	13) 15	18) -70
4) 14	9) 14	14) 9	19) 74
5) 15	10) 0	15) 81	20) -26

Combining like Terms

1) $14x + 6$	9) $6x + 1$	17) $37x + 50$	25) $-29x + 10$
2) $16x - 48$	10) $33x - 31$	18) $19x + 17$	26) $-14x + 14$
3) $11x + 11$	11) $31x - 11$	19) $13x + 19$	27) $-56x + 18$
4) $-24x + 28$	12) $-15x + 20$	20) $22x - 33$	28) $27x - 42$
5) $12x - 5$	13) $13x - 6$	21) $x + 12$	29) $9x + 15$
6) $40x - 13$	14) $-18x - 4$	22) $-25x + 30$	30) $36x - 31$
7) $-8x + 26$	15) $23x + 44$	23) $-42x + 11$	31) $-15x - 1$
8) $-35x + 17$	16) $-10x - 14$	24) $-42x - 27$	32) $51x$

Chapter 6 :

Equations and Inequalities

Topics that you'll practice in this chapter:

- ✓ One–Step Equations
- ✓ Multi–Step Equations
- ✓ Graphing Single–Variable Inequalities
- ✓ One–Step Inequalities
- ✓ Multi-Step Inequalities
- ✓ Systems of Equations
- ✓ Systems of Equations Word Problems

"Life is a math equation. In order to gain the most, you have to know how to convert negatives into positives." – Anonymous

One–Step Equations

🐦 **Find the answer for each equation.**

1) $3x = 90, x =$ ____

2) $5x = 35, x =$ ____

3) $6x = 24, x =$ ____

4) $24x = 144, x =$ ____

5) $x + 15 = 20, x =$ ____

6) $x - 7 = 4, x =$ ____

7) $x - 9 = 2, x =$ ____

8) $x + 15 = 23, x =$ ____

9) $x - 4 = 13, x =$ ____

10) $12 = 16 + x, x =$ ____

11) $x - 10 = 2, x =$ ____

12) $5 - x = -11, x =$ ____

13) $28 = -6 + x, x =$ ____

14) $x - 20 = -35, x =$ ____

15) $x + 14 = -4, x =$ ____

16) $14 = 28 - x, x =$ ____

17) $7 + x = -7, x =$ ____

18) $x - 16 = 4, x =$ ____

19) $30 = x - 15, x =$ ____

20) $x - 5 = -18, x =$ ____

21) $x - 10 = 24, x =$ ____

22) $x - 20 = -25, x =$ ____

23) $x - 17 = 30, x =$ ____

24) $-70 = x - 28, x =$ ____

25) $x - 9 = 13, x =$ ____

26) $36 = 4x, x =$ ____

27) $x - 35 = 25, x =$ ____

28) $x - 25 = 10, x =$ ____

29) $70 - x = 16, x =$ ____

30) $x - 10 = 14, x =$ ____

31) $17 - x = -13, x =$ ____

32) $x - 9 = -30, x =$ ____

Multi–Step Equations

✎ **Find the answer for each equation.**

1) $3x + 3 = 9$

2) $-x + 5 = 12$

3) $4x - 8 = 8$

4) $-(3 - x) = 5$

5) $4x - 8 = 16$

6) $12x - 15 = 9$

7) $2x - 18 = 2$

8) $4x + 8 = 16$

9) $24x + 27 = 75$

10) $-14(3 + x) = 14$

11) $-3(2 + x) = 6$

12) $12 = -(x - 7)$

13) $3(3 - x) = 30$

14) $-15 = -(3x + 6)$

15) $40(3 + x) = 40$

16) $5(x - 10) = 25$

17) $-18 = x + 8x$

18) $3x + 25 = -2x - 10$

19) $7(6 + 3x) = -63$

20) $18 - 3x = -4 - 5x$

21) $4 - 6x = 36 + 2x$

22) $15 + 15x = -5 + 5x$

23) $42 = (-6x) - 7 + 7$

24) $21 = 3x - 21 + 4x$

25) $-18 = -6x - 9 + 3x$

26) $5x - 15 = -29 + 6x$

27) $7x - 18 = 4x + 3$

28) $-7 - 4x = 5(4 - x)$

29) $x - 5 = -5(-3 - x)$

30) $13x - 68 = 15x - 102$

31) $-5x - 3 = -3(9 + 3x)$

32) $-2x - 15 = 6x + 17$

Graphing Single–Variable Inequalities

✎ **Draw a graph for each inequality.**

1) $x > -1$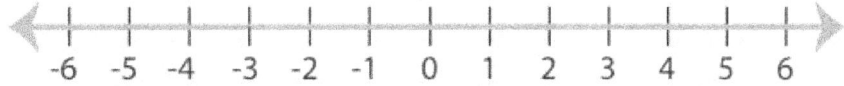

2) $x \leq 2$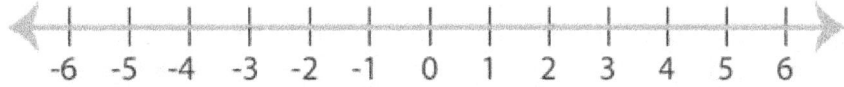

3) $x \geq 0$

4) $x < -3$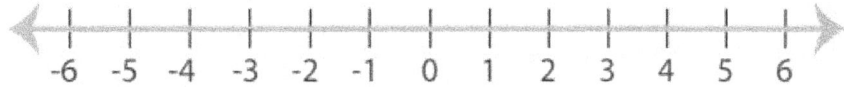

5) $x < \frac{1}{2}$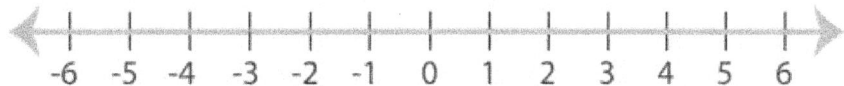

6) $x \leq -2$

7) $x \leq 3$

8) $x \geq -\frac{7}{2}$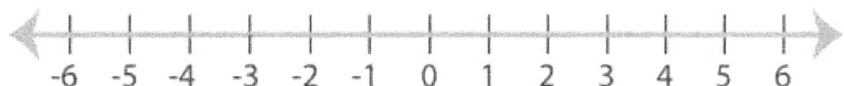

One–Step Inequalities

✎ **Find the answer for each inequality and graph it.**

1) $x + 4 \geq 4$

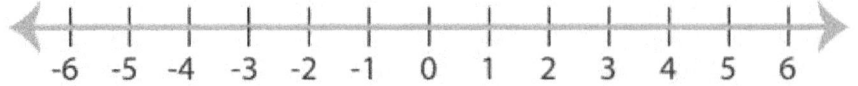

2) $x - 5 \leq 2$

3) $5x > 35$

4) $9 + x \leq 11$

5) $x - 5 < -9$

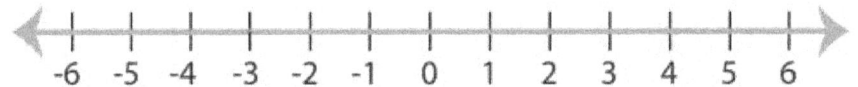

6) $9x \geq 72$

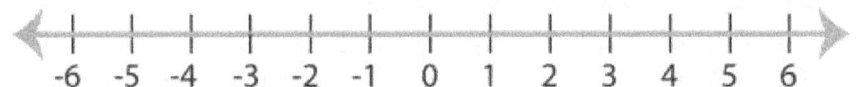

7) $9x \leq 27$

8) $x + 19 > 16$

Multi-Step Inequalities

✍ **Calculate each inequality.**

1) $x - 3 \leq 7$

2) $8 - x \leq 8$

3) $3x - 9 \leq 9$

4) $4x - 4 \geq 8$

5) $x - 7 \geq 1$

6) $5x - 15 \leq 5$

7) $6x - 8 \leq 4$

8) $-11 + 6x \leq 12$

9) $4(x - 4) \leq 16$

10) $3x - 10 \leq 11$

11) $5x - 25 < 25$

12) $9x - 5 < 22$

13) $20 - 7x \geq -15$

14) $33 + 6x < 45$

15) $8 + 8x \geq 96$

16) $7 + 3x < 13$

17) $4x - 3 < 9$

18) $5(2 - 2x) \geq -30$

19) $-(7 + 6x) < 29$

20) $12 - 8x \geq -20$

21) $-4(x - 6) > 24$

22) $\dfrac{3x + 9}{6} \leq 10$

23) $\dfrac{4x - 10}{3} \leq 2$

24) $\dfrac{2x - 8}{3} > 2$

25) $8 + \dfrac{x}{6} < 9$

26) $\dfrac{9x}{7} - 4 < 5$

27) $\dfrac{15x + 45}{15} > 1$

28) $16 + \dfrac{x}{4} < 6$

Systems of Equations

✎ Calculate each system of equations.

1) $-x + y = 2$ $x =$ ___
   $-4x + 2y = 6$ $y =$ ___

2) $-15x + 3y = -9$ $x =$ ___
   $9x - 16y = 48$ $y =$ ___

3) $y = -7$ $x =$ ___
   $6x + 5y = 7$ $y =$ ___

4) $3y = -9x + 15$ $x =$ ___
   $5x - 4y = -3$ $y =$ ___

5) $10x - 9y = -13$ $x =$ ___
   $-5x + 3y = 11$ $y =$ ___

6) $-12x - 16y = 20$ $x =$ ___
   $6x - 12y = 30$ $y =$ ___

7) $5x - 14y = -23$ $x =$ ___
   $-18x + 21y = 24$ $y =$ ___

8) $15x - 21y = -6$ $x =$ ___
   $2x - 3y = -2$ $y =$ ___

9) $-x + 3y = 3$ $x =$ ___
   $-14x + 16y = -10$ $y =$ ___

10) $x + 5y = 50$ $x =$ ___
    $3x + 10y = 80$ $y =$ ___

11) $6x - 7y = -8$ $x =$ ___
    $-x - 4y = -9$ $y =$ ___

12) $2x + 4y = -10$ $x =$ ___
    $2x - 8y = 14$ $y =$ ___

13) $4x + 3y = 12$ $x =$ ___
    $5x - 3y = 15$ $y =$ ___

14) $3x - 2y = 3$ $x =$ ___
    $7x - 8y = 22$ $y =$ ___

15) $3x + 2y = 5$ $x =$ ___
    $-10x - 4y = -14$ $y =$ ___

16) $10x + 7y = 1$ $x =$ ___
    $-5x - 7y = 24$ $y =$ ___

Systems of Equations Word Problems

✍ **Find the answer for each word problem.**

1) Tickets to a movie cost $4 for adults and $3 for students. A group of friends purchased 8 tickets for $31.00. How many adults ticket did they buy? ____

2) At a store, Eva bought two shirts and five hats for $77.00. Nicole bought three same shirts and four same hats for $84.00. What is the price of each shirt? _____

3) A farmhouse shelters 18 animals, some are pigs, and some are ducks. Altogether there are 66 legs. How many pigs are there? _____

4) A class of 214 students went on a field trip. They took 36 vehicles, some cars and some buses. If each car holds 5 students and each bus hold 22 students, how many buses did they take? _____

5) A theater is selling tickets for a performance. Mr. Smith purchased 5 senior tickets and 3 child tickets for $105 for his friends and family. Mr. Jackson purchased 3 senior tickets and 5 child tickets for $79. What is the price of a senior ticket? $_____

6) The difference of two numbers is 10. Their sum is 20. What is the bigger number? $_____

7) The sum of the digits of a certain two–digit number is 7. Reversing its digits increase the number by 9. What is the number? _____

8) The difference of two numbers is 11. Their sum is 25. What are the numbers? _____

9) The length of a rectangle is 5 meters greater than 2 times the width. The perimeter of rectangle is 28 meters. What is the length of the rectangle? _____

10) Jim has 25 nickels and dimes totaling $1.80. How many nickels does he have? _____

Answers of Worksheets

One–Step Equations

1) 30	9) 17	17) −14	25) 22
2) 7	10) −4	18) 20	26) 9
3) 4	11) 12	19) 45	27) 60
4) 6	12) 16	20) −13	28) 35
5) 5	13) 34	21) 34	29) 54
6) 11	14) −15	22) −5	30) 24
7) 11	15) −18	23) 47	31) 30
8) 8	16) 14	24) −42	32) −21

Multi–Step Equations

1) 2	9) 2	17) −2	25) 3
2) −7	10) −4	18) −7	26) 14
3) 4	11) −4	19) −5	27) 7
4) 8	12) −5	20) −11	28) 27
5) 6	13) −7	21) −4	29) −5
6) 2	14) 3	22) −2	30) 17
7) 10	15) −2	23) −7	31) −6
8) 2	16) 15	24) 6	32) −4

Graphing Single–Variable Inequalities

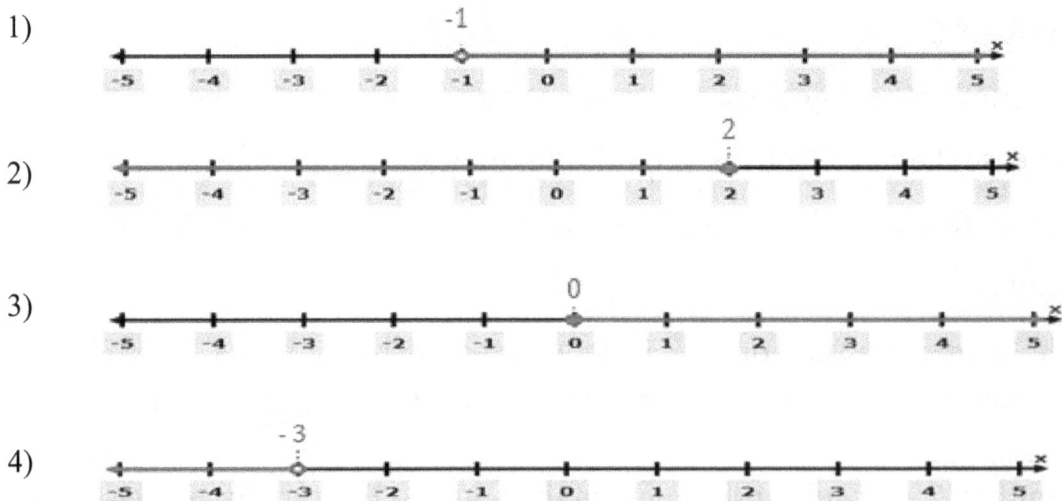

1)

2)

3)

4)

5)

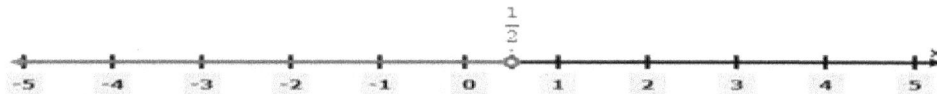

6)

7)

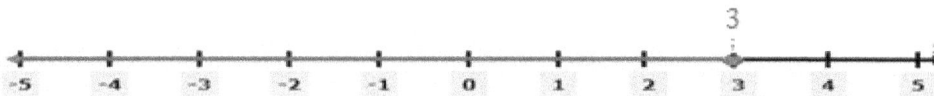

8)

One–Step Inequalities

1)

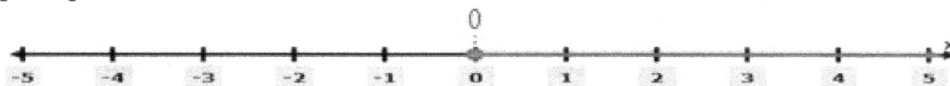

2)

3)

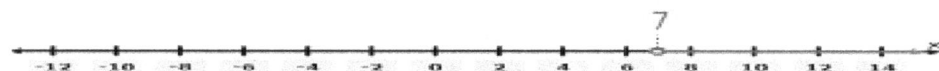

4)

5)

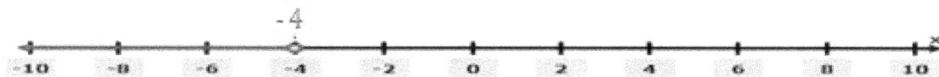

6)

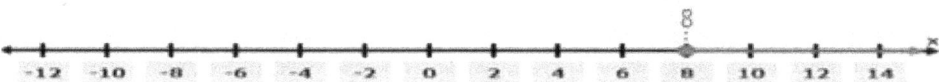

7)

8)

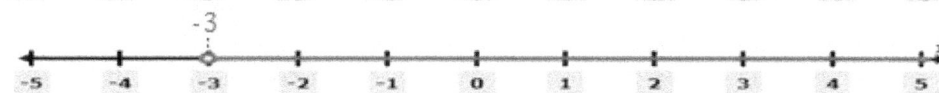

Multi-Step Inequalities

1) $x \leq 10$

2) $x \geq 0$

3) $x \leq 6$

4) $x \geq 3$

5) $x \geq 8$

6) $x \leq 4$

7) $x \leq 2$

8) $x \leq \frac{23}{6}$

9) $x \leq 8$

10) $x \leq 7$

11) $x < 10$

12) $x < 3$

13) $x \leq 5$

14) $x < 2$

15) $x \geq 11$

16) $x < 2$

17) $x < 3$

18) $x \leq 4$

19) $x > -6$

20) $x \leq 4$

21) $x < 0$

22) $x \leq 17$

23) $x \leq 4$

24) $x > 7$

25) $x < 6$ 26) $x < 7$ 27) $x > -2$ 28) $x < -40$

Systems of Equations

1) $x = -1, y = 1$

2) $x = 0, y = -3$

3) $x = 7$

4) $x = 1, y = 2$

5) $x = -4, y = -3$

6) $x = 1, y = -2$

7) $x = 1, y = 2$

8) $x = 8, y = 6$

9) $x = 3, y = 2$

10) $x = -20, y = 14$

11) $x = 1, y = 2$

12) $x = -1, y = -2$

13) $x = 3, y = 0$

14) $x = -2, y = -\frac{9}{2}$

15) $x = 1, y = 1$

16) $x = 5, y = -7$

Systems of Equations Word Problems

1) 7

2) $16

3) 15

4) 2

5) $18

6) 15

7) 34

8) 18, 7

9) 11 meters

10) 14

Chapter 7 :

Linear Functions

Topics that you'll practice in this chapter:

✓ Finding Slope

✓ Graphing Lines Using Line Equation

✓ Writing Linear Equations

✓ Graphing Linear Inequalities

✓ Finding Midpoint

✓ Finding Distance of Two Points

Life is not linear; you have ups and downs. It's how you deal with the troughs that defines you.

Michael Lee-Chin

Finding Slope

✎ **Find the slope of each line.**

1) $y = x + 8$

2) $y = -3x + 5$

3) $y = 2x + 12$

4) $y = -4x + 19$

5) $y = 11 + 6x$

6) $y = 7 - 5x$

7) $y = 8x + 19$

8) $y = -9x + 20$

9) $y = -7x + 4$

10) $y = 3x - 8$

11) $y = \frac{1}{3}x + 8$

12) $y = -\frac{4}{5}x + 9$

13) $-3x + 6y = 30$

14) $4x + 4y = 16$

15) $3y - x = 10$

16) $8y - x = 5$

✎ **Find the slope of the line through each pair of points.**

17) $(2, 3), (7, 10)$

18) $(-3, 5), (2, 15)$

19) $(5, -3), (1, 9)$

20) $(-5, -5), (10, 25)$

21) $(22, 3), (7, 18)$

22) $(-16, 8), (-7, 26)$

23) $(25, 11), (29, 19)$

24) $(26, -19), (14, 17)$

25) $(22, -13), (20, -11)$

26) $(19, 7), (15, -3)$

27) $(5, 7), (11, 19)$

28) $(52, -62), (40, 70)$

Graphing Lines Using Line Equation

✎ **Sketch the graph of each line.**

1) $y = x - 2$ 2) $y = -3x + 2$ 3) $x + y = 0$

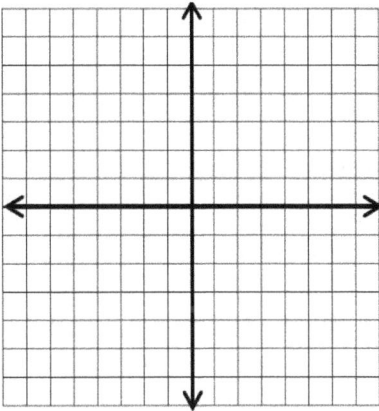   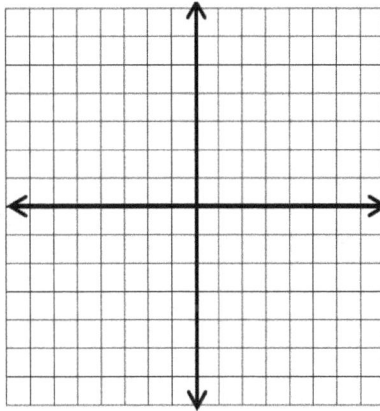   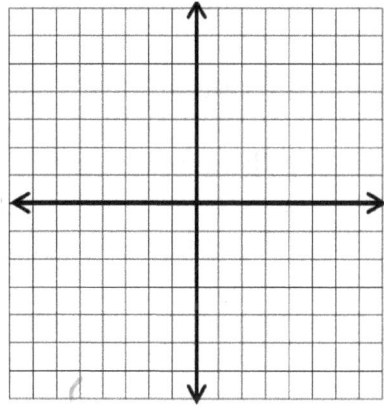

4) $x + y = -3$ 5) $2x + 3y = -4$ 6) $y - 3x + 6 = 0$

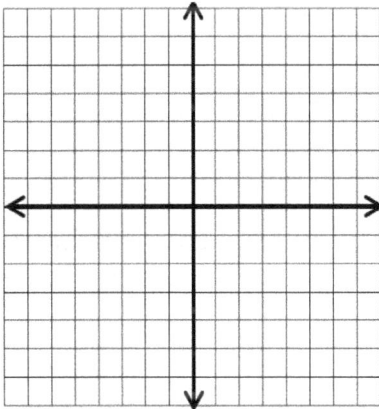   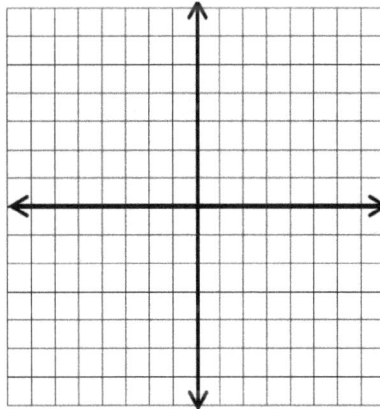   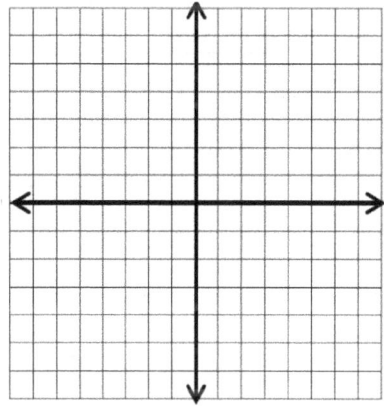

Writing Linear Equations

✎ **Write the equation of the line through the given points.**

1) Through: $(2, -5), (3, 9)$

2) Through: $(-6, 3), (3, 12)$

3) Through: $(10, 7), (5, 27)$

4) Through: $(15, 11), (3, -1)$

5) Through: $(24, 17), (12, -7)$

6) Through: $(8, 29), (4, -7)$

7) Through: $(20, -16), (12, 0)$

8) Through: $(-3, 10), (2, -5)$

9) Through: $(-6, 17), (4, -3)$

10) Through: $(-8, 22), (5, -4)$

11) Through: $(9, 27), (3, -3)$

12) Through: $(11, 32), (9, 4)$

13) Through: $(-3, 13), (-4, 0)$

14) Through: $(-5, 5), (5, 15)$

15) Through: $(18, -32), (11, 3)$

16) Through: $(-4, 25), (4, -15)$

✎ **Find the answer for each problem.**

17) What is the equation of a line with slope 6 and intercept 12?

_____

18) What is the equation of a line with slope -11 and intercept -4?

_____

19) What is the equation of a line with slope -3 and passes through point $(5, 2)$? _____

20) What is the equation of a line with slope -5 and passes through point $(-2, -1)$? _____

21) The slope of a line is -10 and it passes through point $(-3, 0)$. What is the equation of the line? _____

22) The slope of a line is 8 and it passes through point $(0, 7)$. What is the equation of the line? _____

Graphing Linear Inequalities

✎ **Sketch the graph of each linear inequality.**

1) $y > 4x - 5$ 2) $y < 2x + 4$ 3) $y \leq -5x - 2$

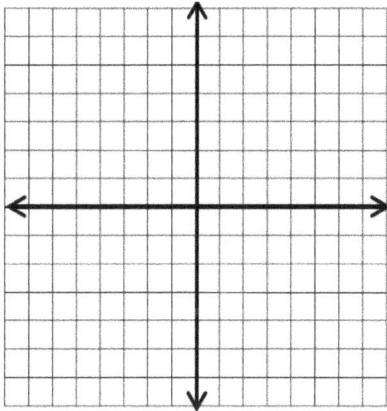 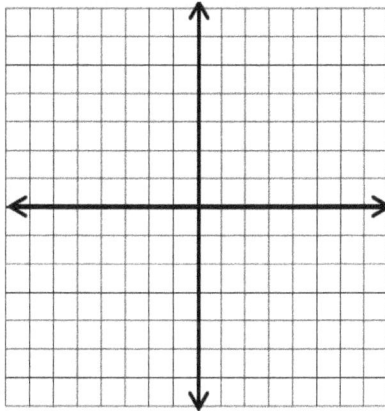 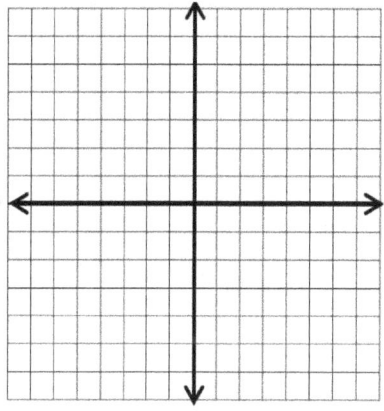

4) $4y \geq 12 + 4x$ 5) $-12y < 3x - 24$ 6) $5y \geq -15x + 10$

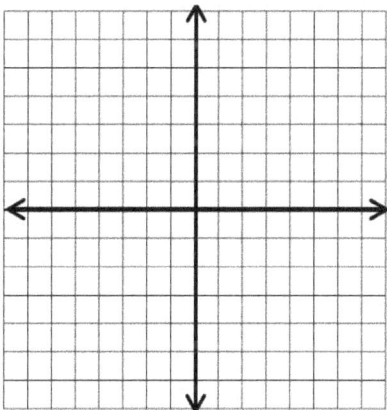 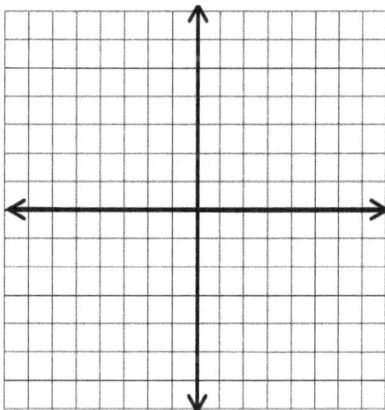 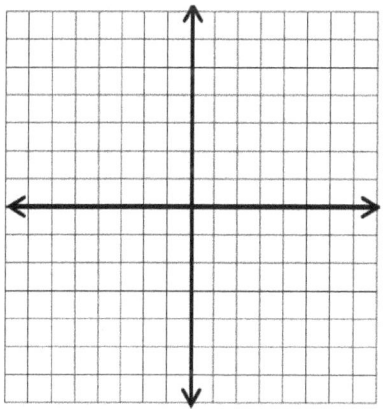

Finding Midpoint

✎ **Find the midpoint of the line segment with the given endpoints.**

1) $(-4, -3), (2, 3)$

2) $(9, 0), (-1, 8)$

3) $(9, -6), (3, 14)$

4) $(-10, -6), (0, 8)$

5) $(2, -5), (14, -15)$

6) $(-10, -3), (4, -13)$

7) $(8, 7), (-8, 13)$

8) $(-3, 6), (-9, 2)$

9) $(-4, 5), (16, -9)$

10) $(7, 14), (9, -2)$

11) $(-8, 6), (6, 6)$

12) $(10, 5), (-2, -3)$

13) $(-5, 12), (-3, 3)$

14) $(12, 7), (8, -2)$

15) $(10, 2), (-6, 14)$

16) $(-1, -2), (-7, 10)$

17) $(7, -7), (13, -13)$

18) $(-3, -8), (11, -4)$

19) $(5, -11), (-8, 9)$

20) $(14, -4), (16, 14)$

21) $(0, -5), (8, -1)$

22) $(3, 0), (-21, 18)$

23) $(17, -3), (-7, -5)$

24) $(26, -12), (6, 24)$

✎ **Find the answer for each problem.**

25) One endpoint of a line segment is $(-3, 7)$ and the midpoint of the line segment is $(-6, 9)$. What is the other endpoint? _____

26) One endpoint of a line segment is $(-3, 7)$ and the midpoint of the line segment is $(1, 5)$. What is the other endpoint? _____

27) One endpoint of a line segment is $(-10, -16)$ and the midpoint of the line segment is $(2, 9)$. What is the other endpoint? _____

Finding Distance of Two Points

✎ **Find the distance between each pair of points.**

1) $(6, 3), (-3, -9)$

2) $(5, 2), (-10, -6)$

3) $(8, 5), (8, 3)$

4) $(-8, -2), (2, 22)$

5) $(6, -7), (-3, -7)$

6) $(12, 0), (-9, -20)$

7) $(3, 20), (3, -5)$

8) $(10, 17), (5, 5)$

9) $(7, -2), (-4, -2)$

10) $(13, 4), (5, -2)$

11) $(11, 13), (5, 5)$

12) $(1, 4), (-23, -3)$

13) $(9, 8), (5, -4)$

14) $(-11, -4), (5, 8)$

15) $(-2, -6), (-2, -12)$

16) $(-1, -4), (23, 3)$

17) $(19, 3), (7, -6)$

18) $(-5, -2), (3, 4)$

19) $(2, 6), (2, -12)$

20) $(-4, -2), (8, -2)$

✎ **Find the answer for each problem.**

21) Triangle ABC is a right triangle on the coordinate system and its vertices are $(-2, 5)$, $(-2, 1)$, and $(1, 1)$. What is the area of triangle ABC? _____

22) Three vertices of a triangle on a coordinate system are $(3, -6)$, $(-5, -12)$, and $(3, -18)$. What is the perimeter of the triangle? _____

23) Four vertices of a rectangle on a coordinate system are $(-2, 2)$, $(-2, 6)$, $(4, 2)$, and $(4, 6)$. What is its perimeter? _____

Answers of Worksheets

Finding Slope

1) 1

2) -3

3) 2

4) -4

5) 6

6) -5

7) 8

8) -9

9) -7

10) 3

11) $\frac{1}{3}$

12) $-\frac{4}{5}$

13) $\frac{1}{2}$

14) -1

15) $\frac{1}{3}$

16) $\frac{1}{8}$

17) $\frac{7}{5}$

18) 2

19) -3

20) 2

21) -1

22) 2

23) 2

24) -3

25) -1

26) $\frac{5}{2}$

27) 2

28) -11

Graphing Lines Using Line Equation

1) $y = x - 2$

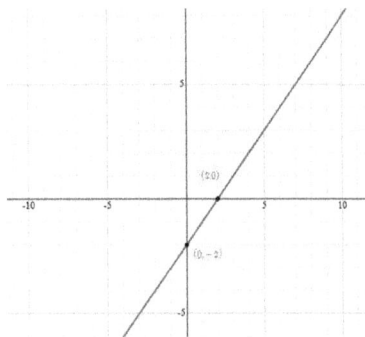

2) $y = -3x + 2$

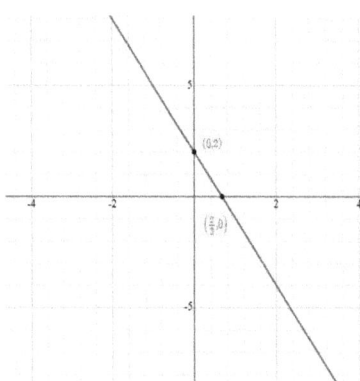

3) $x + y = 0$

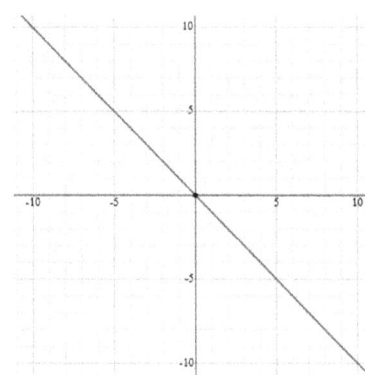

4) $x + y = -3$

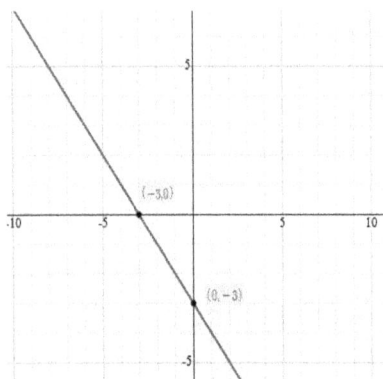

5) $2x + 3y = -4$

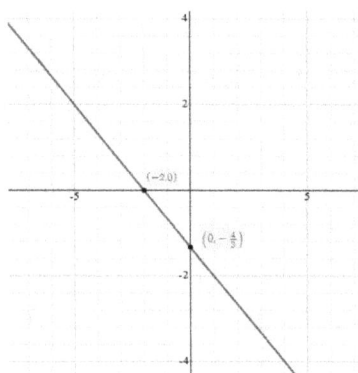

6) $y - 3x + 6 = 0$

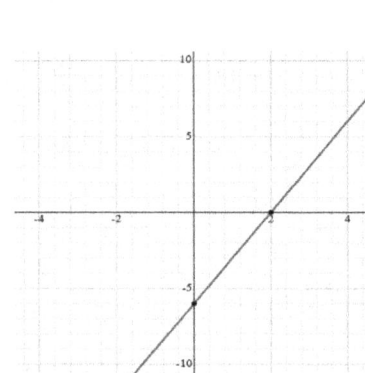

ACT Subject Test – Mathematics

Writing Linear Equations

1) $y = 14x - 33$

2) $y = x + 9$

3) $y = -4x + 47$

4) $y = x - 4$

5) $y = 2x - 31$

6) $y = 9x - 43$

7) $y = -2x + 24$

8) $y = -3x + 1$

9) $y = -2x + 5$

10) $y = -2x + 6$

11) $y = 5x - 18$

12) $y = 14x - 122$

13) $y = 13x + 52$

14) $y = x + 10$

15) $y = -5x + 58$

16) $y = -5x + 5$

17) $y = 6x + 12$

18) $y = -11x - 4$

19) $y = -3x + 17$

20) $y = -5x - 11$

21) $y = -10x - 30$

22) $y = 8x + 7$

Graphing Linear Inequalities

1) $y > 4x - 5$

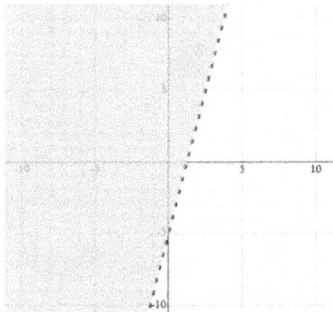

2) $y < 2x + 4$

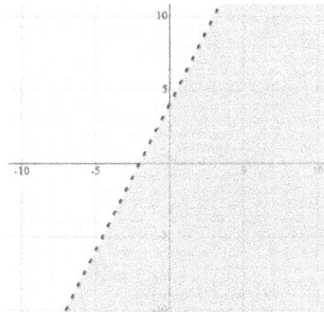

3) $y \leq -5x - 2$

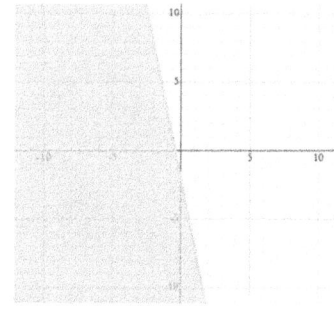

4) $4y \geq 12 + 4x$

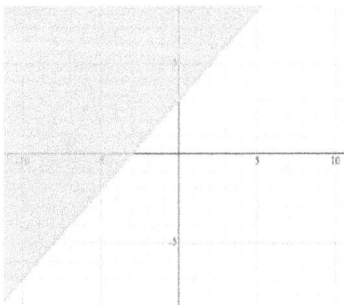

5) $-12y < 3x - 24$

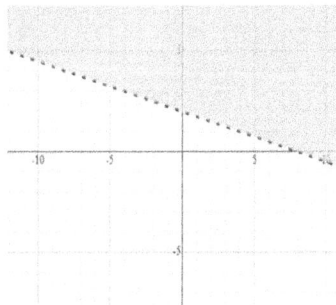

6) $5y \geq -15x + 10$

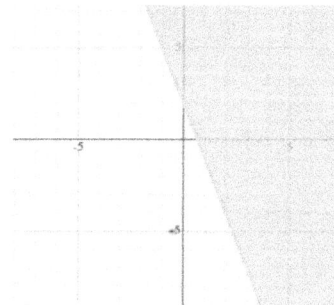

Finding Midpoint

1) $(-1, 0)$

2) $(4, 4)$

3) $(6, 4)$

4) $(-5, 1)$

5) $(8, -10)$

6) $(-3, -8)$

7) $(0, 10)$

8) $(-6, 4)$

9) $(6, -2)$

10) $(8, 6)$

11) $(-1, 6)$

12) $(4, 1)$

13) $(-4, 7.5)$

14) $(10, 2.5)$

15) $(2, 8)$

16) $(-4, 4)$

17) $(10, -10)$

18) $(4, -6)$

19) $(-1.5, -1)$

20) $(15, 5)$

21) $(4, -3)$

22) $(-9, 9)$

23) $(5, -4)$

24) $(16, 6)$

25) $(-9, 11)$

26) $(5, 3)$

27) $(14, 34)$

Finding Distance of Two Points

1) 15

2) 17

3) 2

4) 26

5) 9

6) 29

7) 25

8) 13

9) 11

10) 10

11) 10

12) 25

13) $4\sqrt{10}$

14) 20

15) 6

16) 25

17) 15

18) 10

19) 18

20) 12

21) 6 *square units*

22) 32 *units*

23) 20 *units*

Chapter 8 :

Polynomials

Topics that you'll practice in this chapter:

✓ Writing Polynomials in Standard Form

✓ Simplifying Polynomials

✓ Adding and Subtracting Polynomials

✓ Multiplying Monomials

✓ Multiplying and Dividing Monomials

✓ Multiplying a Polynomial and a Monomial

✓ Multiplying Binomials

✓ Factoring Trinomials

✓ Operations with Polynomials

Mathematics is the supreme judge; from its decisions there is no appeal.

– Tobias Dantzig

Writing Polynomials in Standard Form

✍ **Write each polynomial in standard form.**

1) $11x - 7x =$

2) $-5 + 19x - 19x =$

3) $6x^5 - 12x^3 =$

4) $12 + 17x^4 - 12 =$

5) $5x^2 + 4x - 9x^3 =$

6) $-3x^2 + 12x^5 =$

7) $5x + 8x^3 - 2x^8 =$

8) $-7x^3 + 4x - 9x^6 =$

9) $3x^2 + 22 - 6x =$

10) $3 - 4x + 9x^4 =$

11) $13x^2 + 28x - 8x^3 =$

12) $16 + 4x^2 - 2x^3 =$

13) $19x^2 - 9x + 9x^4 =$

14) $3x^4 - 7x^2 - 2x^3 =$

15) $-51 + 3x^2 - 8x^4 =$

16) $7x^2 - 8x^6 + 4x^4 - 15 =$

17) $6x^4 - 4x^5 + 16 - 3x^3 =$

18) $-2x^6 + 4x - 7x^2 - 5x =$

19) $11x^7 + 8x^5 - 5x^7 - 3x^2 =$

20) $2x^2 - 12x^5 + 8x^2 + 3x^6 =$

21) $4x^5 - 11x^7 - 6x^3 + 16x^5 =$

22) $6x^3 + 3x^5 + 34x^4 - 8x^5 =$

23) $3x(4x + 5 - 2x^2) =$

24) $12x(x^6 + 4x^3) =$

25) $5x(3x^2 + 6x + 4) =$

26) $7x(4 - 2x + 6x^5) =$

27) $3x(4x^4 - 4x^3 + 2) =$

28) $4x(2x^5 + 6x^2 - 3) =$

29) $5x(3x^4 + 4x^3 + 2x) =$

30) $2x(3x - 2x^3 + 4x^6) =$

Simplifying Polynomials

✎ **Simplify each expression.**

1) $3(4x - 20) =$

2) $5x(3x - 4) =$

3) $6x(5x - 7) =$

4) $3x(7x + 5) =$

5) $5x(4x - 3) =$

6) $6x(8x + 2) =$

7) $(3x - 2)(x - 4) =$

8) $(x - 5)(2x + 6) =$

9) $(x - 3)(x - 7) =$

10) $(3x + 4)(3x - 4) =$

11) $(5x - 4)(5x - 2) =$

12) $6x^2 + 6x^2 - 8x^4 =$

13) $3x - 2x^2 + 5x^3 + 7 =$

14) $7x + 4x^2 - 10x^3 =$

15) $12x^2 + 5x^5 - 6x^3 =$

16) $-5x^2 + 4x^6 + 6x^8 =$

17) $-12x^3 + 10x^5 - 4x^6 + 4x =$

18) $11 - 7x^2 + 4x^2 - 16x^3 + 11 =$

19) $2x^2 - 9x + 4x^3 + 15x - 10x =$

20) $13 - 7x^5 + 6x^5 - 4x^2 + 5 =$

21) $-5x^8 + x^6 - 14x^3 + 5x^8 =$

22) $(7x^4 - 4) + (7x^4 - 2x^4) =$

23) $3(3x^4 - 4x^3 - 6x^4) =$

24) $-5(x^9 + 8) - 5(10 - x^9) =$

25) $8x^3 - 9x^4 - 2x + 19 - 8x^3 =$

26) $11 - 8x^3 + 6x^3 - 7x^5 + 6 =$

27) $(5x^3 - 4x) - (6x - 2 - 6x^3) =$

28) $4x^2 - 5x^4 - x(3x^3 + 2x) =$

29) $6x + 6x^5 - 10 - 4(x^5 - 3) =$

30) $4 - 3x^4 + (6x^5 - 2x^4 + 5x^5) =$

31) $-(x^5 + 4) - 8(3 + x^5) =$

32) $(4x^3 - 3x) - (3x - 5x^3) =$

Adding and Subtracting Polynomials

✎ **Add or subtract expressions.**

1) $(-2x^2 - 3) + (3x^2 + 4) =$

2) $(4x^3 + 6) - (7 - 2x^3) =$

3) $(4x^5 + 5x^2) - (2x^5 + 15) =$

4) $(6x^3 - 2x^2) + (5x^2 - 4x) =$

5) $(10x^4 + 28x) - (34x^4 + 6) =$

6) $(7x^2 - 3) + (7x^2 + 3) =$

7) $(9x^2 + 4) - (10 - 5x^2) =$

8) $(6x^2 + x^5) - (x^5 + 4) =$

9) $(4x^3 - x) + (3x - 7x^3) =$

10) $(11x + 10) - (8x + 10) =$

11) $(15x^3 - 3x) - (3x - 4x^3) =$

12) $(4x - x^5) - (6x^5 + 8x) =$

13) $(2x^2 - 7x^7) - (4x^7 - 6x) =$

14) $(3x^2 - 5) + (8x^2 + 4x^5) =$

15) $(9x^4 + 5x^5) - (x^5 - 9x^4) =$

16) $(-4x^3 - 2x) + (9x - 5x^3) =$

17) $(4x - 3x^2) - (148x^2 + x) =$

18) $(5x - 8x^4) - (3x^4 - 4x^2) =$

19) $(8x^4 - 4) + (2x^4 - 3x^2) =$

20) $(5x^6 + 7x^3) - (x^3 - 5x^6) =$

21) $(-2x^2 + 20x^5 + 5x^4) + (12x^4 + 8x^5 + 24x^2) =$

22) $(7x^4 - 9x^7 - 6x) - (-3x^4 - 9x^7 + 6x) =$

23) $(14x + 12x^4 - 18x^6) + (20x^4 + 18x^6 - 10x) =$

24) $(5x^8 - 6x^6 - 4x) - (5x^3 + 9x^6 - 7x) =$

25) $(11x^2 - 6x^4 - 3x) - (-4x^2 - 12x^4 + 9x) =$

26) $(-5x^9 + 14x^3 + 3x^7) + (10x^7 + 26x^3 + 3x^9) =$

Multiplying Monomials

🖎 **Simplify each expression.**

1) $6u^8 \times (-u^2) =$

2) $(-5p^8) \times (-2p^3) =$

3) $4xy^3z^5 \times 3z^4 =$

4) $3u^5t \times 8ut^4 =$

5) $(-5a^2) \times (-7a^3b^6) =$

6) $-3a^4b^3 \times 6a^2b =$

7) $13xy^5 \times x^4y^4 =$

8) $6p^4q^3 \times (-8pq^6) =$

9) $8s^4t^3 \times 4st^3 =$

10) $(-6x^4y^3) \times 6x^2y =$

11) $3xy^7z \times 12z^3 =$

12) $24xy \times x^2y =$

13) $13pq^4 \times (-3p^2q) =$

14) $13s^3t^4 \times st^4 =$

15) $11p^5 \times (-6p^3) =$

16) $(-8p^3q^5r) \times 3pq^4r^6 =$

17) $(-4a^4) \times (-7a^3b) =$

18) $6u^6v^2 \times (-5u^3v^4) =$

19) $9u^5 \times (-3u) =$

20) $-6xy^5 \times 4x^2y =$

21) $13y^5z^3 \times (-y^3z) =$

22) $8a^4bc^3 \times 2abc^3 =$

23) $(-7p^5q^6) \times (-5p^4q^2) =$

24) $4u^5v^3 \times (-4u^7v^3) =$

25) $17y^4z^5 \times (-y^6z) =$

26) $(-5pq^3r^2) \times 8p^2q^4r =$

27) $3ab^5c^6 \times 5a^4bc^2 =$

28) $6x^3yz^2 \times 3x^2y^7z^3 =$

Multiplying and Dividing Monomials

✎ **Simplify each expression.**

1) $(5x^5)(2x^2) =$

2) $(4x^4)(6x^2) =$

3) $(3x^4)(7x^4) =$

4) $(5x^6)(4x^2) =$

5) $(12x^4)(3x^6) =$

6) $(4yx^8)(8y^4x^3) =$

7) $(14x^4y)(x^3y^5) =$

8) $(-5x^3y^4)(2x^3y^5) =$

9) $(-6x^4y^2)(-3x^3y^5) =$

10) $(5x^3y)(-5x^2y^3) =$

11) $(6x^4y^3)(4x^3y^4) =$

12) $(4x^3y^2)(5x^2y^4) =$

13) $(12x^3y^6)(4x^4y^{10}) =$

14) $(15x^3y^5)(3x^4y^6) =$

15) $(7x^2y^7)(8x^6y^7) =$

16) $(-3x^3y^8)(7x^9y^4) =$

17) $\dfrac{5x^6y^6}{xy^4} =$

18) $\dfrac{19x^7y^5}{19x^6y} =$

19) $\dfrac{56x^4y^4}{8xy} =$

20) $\dfrac{81x^5y^6}{9x^4y^5} =$

21) $\dfrac{36x^7y^6}{9x^2y^3} =$

22) $\dfrac{48x^9y^7}{4x^4y^6} =$

23) $\dfrac{88x^{18}y^{12}}{11x^8y^9} =$

24) $\dfrac{30x^7y^6}{6x^8y^3} =$

25) $\dfrac{150x^7y^6}{30x^4y^6} =$

26) $\dfrac{-42x^{18}y^{14}}{6x^4y^9} =$

27) $\dfrac{-36x^7y^8}{9x^5y^8} =$

Multiplying a Polynomial and a Monomial

🖎 **Find each product.**

1) $x(2x + 4) =$

2) $6(4 - 2x) =$

3) $5x(4x + 2) =$

4) $x(-4x + 5) =$

5) $8x(2x - 2) =$

6) $6(2x - 4y) =$

7) $7x(5x - 5) =$

8) $3x(12x + 2y) =$

9) $4x(x + 6y) =$

10) $11x(3x + 4y) =$

11) $7x(3x + 2) =$

12) $10x(4x - 10y) =$

13) $9x(3x - 2y) =$

14) $7x(x - 4y + 6) =$

15) $8x(2x^2 + 5y^2) =$

16) $12x(2x + 3y) =$

17) $4(2x^4 - 4y^4) =$

18) $4x(-3x^2y + 4y) =$

19) $-4(5x^3 - 2xy + 4) =$

20) $4(x^2 - 5xy - 6) =$

21) $8x(2x^3 - 5xy + 2x) =$

22) $-6x(-2x^3 - 6x + 2xy) =$

23) $3(2x^2 + xy - 9y^2) =$

24) $4x(5x^3 - 3x + 7) =$

25) $6(3x^{22} - 2x - 5) =$

26) $x^2(-2x^3 + 4x + 3) =$

27) $x^2(4x^3 + 10 - 2x) =$

28) $4x^4(3x^3 - 2x + 5) =$

29) $2x^2(4x^4 - 5xy + 7y^3) =$

30) $5x^2(5x^4 - 3x + 9) =$

31) $7x^2(6x^2 + 3x - 6) =$

32) $4x(x^3 - 4xy + 2y^2) =$

Multiplying Binomials

✍ **Find each product.**

1) $(x + 3)(x + 6) =$

2) $(x - 4)(x + 3) =$

3) $(x - 3)(x - 8) =$

4) $(x + 8)(x + 9) =$

5) $(x - 2)(x - 12) =$

6) $(x + 5)(x + 5) =$

7) $(x - 6)(x + 7) =$

8) $(x - 8)(x - 3) =$

9) $(x + 7)(x + 12) =$

10) $(x - 4)(x + 8) =$

11) $(x + 8)(x + 8) =$

12) $(x + 2)(x + 7) =$

13) $(x - 6)(x + 6) =$

14) $(x - 5)(x + 5) =$

15) $(x + 11)(x + 11) =$

16) $(x + 6)(x + 9) =$

17) $(x - 2)(x + 2) =$

18) $(x - 4)(x + 7) =$

19) $(3x + 5)(x + 6) =$

20) $(5x - 6)(4x + 8) =$

21) $(x - 7)(3x + 7) =$

22) $(x - 9)(x - 4) =$

23) $(x - 12)(x + 2) =$

24) $(2x - 4)(5x + 4) =$

25) $(3x - 8)(x + 8) =$

26) $(7x - 2)(6x + 3) =$

27) $(4x + 5)(3x + 5) =$

28) $(7x - 4)(9x + 4) =$

29) $(x + 2)(2x - 8) =$

30) $(5x - 4)(5x + 4) =$

31) $(3x + 2)(3x - 7) =$

32) $(x^2 + 8)(x^2 - 8) =$

Factoring Trinomials

✎ **Factor each trinomial.**

1) $x^2 + 8x + 12 =$

2) $x^2 - 6x + 5 =$

3) $x^2 + 15x + 36 =$

4) $x^2 - 12x + 35 =$

5) $x^2 - 11x + 18 =$

6) $x^2 - 9x + 18 =$

7) $x^2 + 18x + 72 =$

8) $x^2 - x - 72 =$

9) $x^2 + 4x - 21 =$

10) $x^2 - 13x + 22 =$

11) $x^2 + 2x - 24 =$

12) $x^2 - 3x - 40 =$

13) $x^2 - 3x - 70 =$

14) $x^2 + 26x + 169 =$

15) $4x^2 - 7x - 15 =$

16) $x^2 - 14x + 33 =$

17) $10x^2 + 5x - 15 =$

18) $6x^2 - 4x - 42 =$

19) $x^2 + 12x + 36 =$

20) $5x^2 + 17x - 12 =$

✎ **Calculate each problem.**

21) The area of a rectangle is $x^2 - x - 56$. If the width of rectangle is $x + 7$, what is its length? _____

22) The area of a parallelogram is $4x^2 + 17x - 15$ and its height is $x + 5$. What is the base of the parallelogram? _____

23) The area of a rectangle is $6x^2 - 22x + 12$. If the width of the rectangle is $3x - 2$, what is its length? _____

Operations with Polynomials

✎ **Find each product.**

1) $4(5x + 3) = $ _____

2) $8(2x + 6) = $ _____

3) $2(5x - 2) = $ _____

4) $-4(7x - 3) = $ _____

5) $3x^2(9x + 1) = $ _____

6) $4x^6(7x - 9) = $ _____

7) $3x^4(-7x + 3) = $ _____

8) $-8x^4(5x - 8) = $ _____

9) $7(x^2 + 5x - 3) = $ _____

10) $9(5x^2 - 7x + 5) = $ _____

11) $3(3x^2 + 3x + 2) = $ _____

12) $5x(3x^2 + 5x + 8) = $ _____

13) $(5x + 7)(3x - 3) = $ _____

14) $(9x + 3)(3x - 5) = $ _____

15) $(6x + 3)(4x - 2) = $ _____

16) $(7x - 2)(3x + 5) = $ _____

✎ **Calculate each problem.**

17) The measures of two sides of a triangle are $(2x + 5y)$ and $(6x - 3y)$. If the perimeter of the triangle is $(13x + 4y)$, what is the measure of the third side? _____

18) The height of a triangle is $(8x + 5)$ and its base is $(4x - 3)$. What is the area of the triangle? _____

19) One side of a square is $(6x + 2)$. What is the area of the square? _____

20) The length of a rectangle is $(5x - 8y)$ and its width is $(15x + 8y)$. What is the perimeter of the rectangle? _____

21) The side of a cube measures $(x + 2)$. What is the volume of the cube? _____

22) If the perimeter of a rectangle is $(28x + 6y)$ and its width is $(5x + 2y)$, what is the length of the rectangle? _____

Answers of Worksheets

Writing Polynomials in Standard Form

1) $4x$

2) -5

3) $6x^5 - 12x^3$

4) $14x^4$

5) $-9x^3 + 5x^2 + 4x$

6) $12x^5 - 3x^2$

7) $-2x^8 + 8x^3 + 5x$

8) $-9x^6 - 7x^3 + 4x$

9) $3x^2 - 6x + 22$

10) $9x^4 - 4x + 3$

11) $-8x^3 + 13x^2 + 28x$

12) $-2x^3 + 4x^2 + 16$

13) $9x^4 + 19x^2 - 9x$

14) $3x^4 - 2x^3 - 7x^2$

15) $-8x^4 + 3x^2 - 51$

16) $-8x^6 + 4x^4 + 7x^2 - 15$

17) $-4x^5 + 6x^4 - 3x^3 + 16$

18) $-2x^6 - 7x^2 - x$

19) $6x^7 + 8x^5 - 3x^2$

20) $3x^6 - 12x^5 + 10x^2$

21) $-11x^7 + 20x^5 - 6x^3$

22) $-5x^5 + 34x^4 + 6x^3$

23) $-6x^3 + 12x^2 + 15x$

24) $12x^7 + 48x^4$

25) $15x^3 + 30x^2 + 20x$

26) $42x^6 - 14x^2 + 28x$

27) $12x^5 - 12x^4 + 6x$

28) $8x^6 + 24x^3 - 12x$

29) $15x^5 + 20x^4 + 10x^2$

30) $8x^7 - 4x^4 + 6x^2$

Simplifying Polynomials

1) $12x - 60$

2) $15x^2 - 20x$

3) $30x^2 - 42x$

4) $21x^2 + 15x$

5) $20x^2 - 15x$

6) $48x^2 + 12x$

7) $3x^2 - 14x + 8$

8) $2x^2 - 4x - 30$

9) $x^2 - 10x + 21$

10) $9x^2 - 16$

11) $25x^2 - 30x + 8$

12) $-8x^4 + 12x^2$

13) $5x^3 - 2x^2 + 3x + 7$

14) $-10x^3 + 4x^2 + 7x$

15) $5x^5 - 6x^3 + 12x^2$

16) $6x^8 + 4x^6 - 5x^2$

17) $-4x^6 + 10x^5 - 12x^3 + 4x$

18) $-16x^3 - 3x^2 + 22$

19) $4x^3 + 2x^2 - 4x$

20) $-x^5 - 4x^2 + 18$

21) $x^6 - 14x^3$

22) $12x^4 - 4$

23) $-9x^4 - 12x^3$

24) -90

25) $-9x^4 - 2x + 19$ 29) $2x^5 + 6x + 2$

26) $-7x^5 - 2x^3 + 17$ 30) $11x^5 - 5x^4 + 4$

27) $11x^3 - 10x + 2$ 31) $-9x^5 - 28$

28) $-8x^4 + 2x^2$ 32) $9x^3 - 6x$

Adding and Subtracting Polynomials

1) $x^2 + 1$ 10) $3x$ 19) $10x^4 - 3x^2 - 4$

2) $6x^3 - 1$ 11) $19x^3 - 6x$ 20) $10x^6 + 6x^3$

3) $2x^5 + 5x^2 - 15$ 12) $-7x^5 - 4x$ 21) $28x^5 + 17x^4 + 22x^2$

4) $6x^3 + 3x^2 - 4x$ 13) $-11x^7 + 2x^2 + 6x$ 22) $10x^4 - 12x$

5) $-24x^4 + 28x - 6$ 14) $4x^5 + 11x^2 - 5$ 23) $32x^4 + 4x$

6) $14x^2$ 15) $4x^5 + 18x^4$ 24) $5x^8 - 15x^6 - 5x^3 + 3x$

7) $14x^2 - 6$ 16) $-9x^3 + 7x$ 25) $6x^4 + 15x^2 - 12x$

8) $6x^2 - 4$ 17) $-151x^2 + 3x$ 26) $-2x^9 + 13x^7 + 40x^3$

9) $-3x^3 + 2x$ 18) $-11x^4 + 4x^2 + 5x$

Multiplying Monomials

1) $-6u^{10}$ 11) $36xy^7z^4$ 21) $-13y^8z^4$

2) $10p^{11}$ 12) $24px^3y^2$ 22) $16a^5b^2c^6$

3) $12xy^3z^9$ 13) $-39p^3q^5$ 23) $35p^9q^8$

4) $24u^6t^5$ 14) $13s^4t^8$ 24) $-16u^{12}v^6$

5) $35a^5b^6$ 15) $-66p^8$ 25) $-17y^{10}z^6$

6) $-18a^6b^4$ 16) $-24p^4q^9r^7$ 26) $-40p^3q^7r^3$

7) $13x^5y^9$ 17) $28a^7b$ 27) $15a^5b^6c^8$

8) $-48p^5q^9$ 18) $-30u^9v^6$ 28) $18x^5y^8z^5$

9) $32s^5t^6$ 19) $-27u^6$

10) $-36x^6y^4$ 20) $-24x^3y^6$

Multiplying and Dividing Monomials

1) $10x^7$ 5) $36x^{10}$ 9) $18x^7y^7$

2) $24x^6$ 6) $32x^{11}y^5$ 10) $-25x^5y^4$

3) $21x^8$ 7) $14x^7y^6$ 11) $24x^7y^7$

4) $20x^8$ 8) $-10x^6y^9$ 12) $20x^5y^6$

13) $48x^7y^{16}$

14) $45x^7y^{11}$

15) $56x^8y^{14}$

16) $-21x^{12}y^{12}$

17) $5x^5y^2$

18) xy^4

19) $7x^3y^3$

20) $9xy$

21) $4x^5y^3$

22) $12x^5y$

23) $8x^{10}y^3$

24) $5x^{-1}y^3$

25) $5x^3$

26) $-7x^{14}y^5$

27) $-4x^2$

Multiplying a Polynomial and a Monomial

1) $2x^2 + 4x$

2) $-12x + 24$

3) $20x^2 + 10x$

4) $-4x^2 + 5x$

5) $16x^2 - 16x$

6) $12x - 24y$

7) $35x^2 - 35x$

8) $36x^2 + 6xy$

9) $4x^2 + 24xy$

10) $33x^2 + 44xy$

11) $21x^2 + 14x$

12) $40x^2 - 100xy$

13) $27x^2 - 18xy$

14) $7x^2 - 28xy + 42x$

15) $16x^3 + 40xy^2$

16) $24x^2 + 36xy$

17) $8x^4 - 16y^4$

18) $-12x^3y + 16xy$

19) $-20x^3 + 8xy - 16$

20) $4x^2 - 20xy - 24$

21) $16x^4 - 40x^2y + 16x^2$

22) $12x^4 + 36x^2 - 12x^2y$

23) $6x^2 + 3xy - 27y^2$

24) $20x^4 - 12x^2 + 28x$

25) $18x^{22} - 12x - 30$

26) $-2x^5 + 4x^3 + 3x^2$

27) $4x^5 - 2x^3 + 10x^2$

28) $12x^7 - 8x^5 + 20x^4$

29) $8x^6 - 10x^3y + 14x^2y^3$

30) $25x^6 - 15x^3 + 45x^2$

31) $42x^4 + 21x^3 - 42x^2$

32) $4x^4 - 16x^2y + 8xy^2$

Multiplying Binomials

1) $x^2 + 9x + 18$

2) $x^2 - x - 12$

3) $x^2 - 11x + 24$

4) $x^2 + 17x + 72$

5) $x^2 - 14x + 24$

6) $x^2 + 10x + 25$

7) $x^2 + x - 42$

8) $x^2 - 11x + 24$

9) $x^2 + 19x + 84$

10) $x^2 + 4x - 32$

11) $x^2 + 16x + 64$

12) $x^2 + 9x + 14$

13) $x^2 - 36$

14) $x^2 - 25$

15) $x^2 + 22x + 121$

16) $x^2 + 15x + 54$

17) $x^2 - 4$

18) $x^2 + 3x - 28$

19) $3x^2 + 23x + 30$

20) $20x^2 + 16x - 48$

21) $3x^2 - 14x - 49$

22) $x^2 - 13x + 36$

23) $x^2 - 10x - 24$

24) $10x^2 - 12x - 16$

25) $3x^2 + 16x - 64$

26) $42x^2 + 9x - 6$

27) $12x^2 + 35x + 25$

28) $63x^2 - 8x - 16$

29) $2x^2 - 4x - 16$

30) $25x^2 - 16$

31) $9x^2 - 15x - 14$

32) $x^4 - 64$

Factoring Trinomials

1) $(x + 6)(x + 2)$

2) $(x - 5)(x - 1)$

3) $(x + 12)(x + 3)$

4) $(x - 5)(x - 7)$

5) $(x - 2)(x - 9)$

6) $(x - 6)(x - 3)$

7) $(x + 6)(x + 12)$

8) $(x + 8)(x - 9)$

9) $(x - 3)(x + 7)$

10) $(x - 11)(x - 2)$

11) $(x - 4)(x + 6)$

12) $(x - 8)(x + 5)$

13) $(x + 7)(x - 10)$

14) $(x + 13)(x + 13)$

15) $(4x + 5)(x - 3)$

16) $(x - 11)(x - 3)$

17) $(5x - 5)(2x + 3)$

18) $(2x - 6)(3x + 7)$

19) $(x + 6)(x + 6)$

20) $(5x - 3)(x + 4)$

21) $(x - 8)$

22) $(4x - 3)$

23) $(2x - 6)$

Operations with Polynomials

1) $20x + 12$

2) $16x + 48$

3) $10x - 4$

4) $-28x + 12$

5) $27x^3 + 3x^2$

6) $28x^7 - 36x^6$

7) $-21x^5 + 9x^4$

8) $-40x^5 + 64x^4$

9) $7x^2 + 35x - 21$

10) $45x^2 - 63x + 45$

11) $9x^2 + 9x + 6$

12) $15x^3 + 25x^2 + 40x$

13) $15x^2 + 6x - 21$

14) $27x^2 - 36x - 15$

15) $24x^2 - 6$

16) $21x^2 + 29x - 10$

17) $(5x + 2y)$

18) $16x^2 - 2x - \frac{15}{2}$

19) $36x^2 + 24x + 4$

20) $40x$

21) $x^3 + 6x^2 + 12x + 8$

22) $(9x + y)$

Chapter 9 :

Complex Numbers

Topics that you'll practice in this chapter:

- ✓ Adding and Subtracting Complex Numbers
- ✓ Multiplying and Dividing Complex Numbers
- ✓ Graphing Complex Numbers
- ✓ Rationalizing Imaginary Denominators

Mathematics is a hard thing to love. It has the unfortunate habit, like a rude dog, of turning its most unfavorable side towards you when you first make contact with it. — *David Whiteland*

Adding and Subtracting Complex Numbers

✎ **Simplify.**

1) $(7i) - (3i) =$

2) $(5i) + (4i) =$

3) $(2i) + (8i) =$

4) $(-8i) - (3i) =$

5) $(14i) + (6i) =$

6) $(6i) - (-10i) =$

7) $(-2i) + (-5i) =$

8) $(13i) - (5i) =$

9) $(-22i) - (11i) =$

10) $(-4i) + (2 + 6i) =$

11) $(10 - 5i) + (-3i) =$

12) $(-8i) + (6 + 12i) =$

13) $1 + (5 - 4i) =$

14) $(13i) - (-8 + 2i) =$

15) $(8 + 12i) - (-10i) =$

16) $(10 + i) + (-5i) =$

17) $(11i) - (-7i + 9) =$

18) $(10i + 12) + (-2i) =$

19) $(20) - (16 + 4i) =$

20) $(3 + 3i) + (8 + 4i) =$

21) $(15 - 7i) + (3 + 4i) =$

22) $(12 + 6i) + (10 + 17i) =$

23) $(-5 + 6i) - (-16 - 12i) =$

24) $(-4 + 14i) - (-9 + 11i) =$

25) $(-22 + 4i) - (-7 - 22i) =$

26) $(-26 - 18i) + (3 + 34i) =$

27) $(-19 - 13i) - (-7 - 20i) =$

28) $-21 + (5i) + (-32 + 14i) =$

29) $30 - (7i) + (3 - 11i) =$

30) $28 + (-32 - 10i) - 7 =$

31) $(-44i) + (2 - 7i) + 9 =$

32) $(-21i) - (12 - 9i) + 21i =$

Multiplying and Dividing Complex Numbers

✎ **Simplify.**

1) $(5i)(-3i) =$

2) $(-8i)(2i) =$

3) $(3i)(-3i)(-3i) =$

4) $(6i)(-6i) =$

5) $(-3 - 4i)(2 + i) =$

6) $(5 - 2i)^2 =$

7) $(5 - 2i)(6 - 4i) =$

8) $(1 + 6i)^2 =$

9) $(5i)(-3i)(2 - 4i) =$

10) $(11 - 2i)(2 - 4i) =$

11) $(-3 + i)(6 + 5i) =$

12) $(2 - 8i)(6 - 4i) =$

13) $3(4i) - (6i)(-2 + 5i) =$

14) $\dfrac{5}{-25i} =$

15) $\dfrac{3-4i}{-5i} =$

16) $\dfrac{6+12i}{2i} =$

17) $\dfrac{20i}{-5+4i} =$

18) $\dfrac{-6-9i}{4i} =$

19) $\dfrac{4i}{8-2i} =$

20) $\dfrac{4-7i}{6-2i} =$

21) $\dfrac{3-2i}{-1-1i} =$

22) $\dfrac{-5-5i}{-4-i} =$

23) $\dfrac{-6+2i}{-10-4i} =$

24) $\dfrac{-8-4i}{-2+4i} =$

25) $\dfrac{2+3i}{1-4i} =$

Graphing Complex Numbers

✎ **Identify each complex number graphed.**

1)

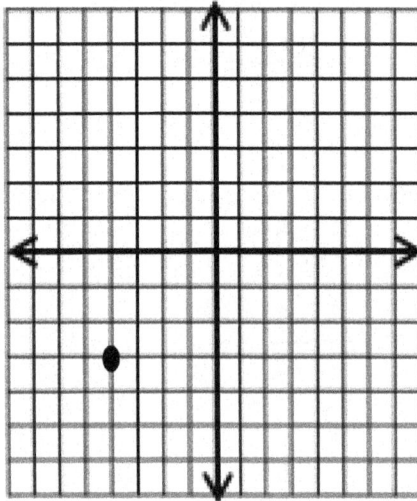

2)

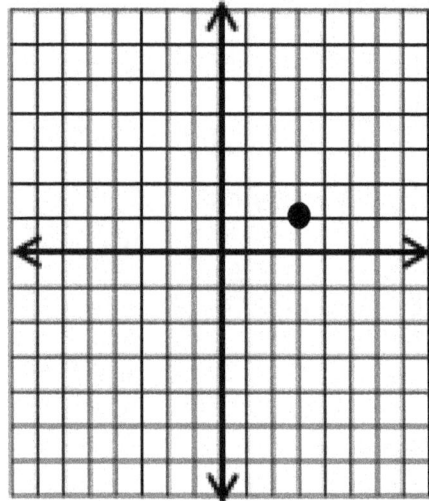

3)

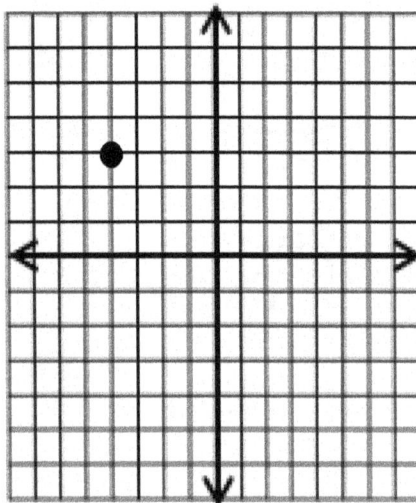

4)

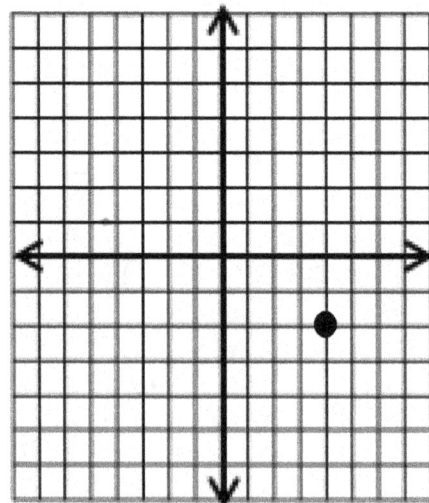

Rationalizing Imaginary Denominators

✏️ **Simplify.**

1) $\dfrac{-7}{-7i} =$

2) $\dfrac{-3}{-15i} =$

3) $\dfrac{-3}{-39i} =$

4) $\dfrac{24}{-3i} =$

5) $\dfrac{5}{2i} =$

6) $\dfrac{16}{-4i} =$

7) $\dfrac{14}{-6i} =$

8) $\dfrac{-17}{3i} =$

9) $\dfrac{4x}{5yi} =$

10) $\dfrac{10-10i}{-2i} =$

11) $\dfrac{5-11i}{-i} =$

12) $\dfrac{21+4i}{4i} =$

13) $\dfrac{8i}{-1+4i} =$

14) $\dfrac{10i}{-6+8i} =$

15) $\dfrac{-25-5i}{-5+5i} =$

16) $\dfrac{-7-2i}{3+1i} =$

17) $\dfrac{-12-6i}{8-6i} =$

18) $\dfrac{-14+7i}{-7i} =$

19) $\dfrac{12+3i}{3i} =$

20) $\dfrac{-2-i}{4-3i} =$

21) $\dfrac{-11+4i}{-5i} =$

22) $\dfrac{8+2i}{-5-2i} =$

23) $\dfrac{-9-5i}{-8-2i} =$

24) $\dfrac{4i-1}{-5-2i} =$

Answers of Worksheets – Chapter 9

Adding and Subtracting Complex Numbers

1) $4i$
2) $9i$
3) $10i$
4) $-11i$
5) $20i$
6) $16i$
7) $-7i$
8) $8i$

9) $-33i$
10) $2 + 2i$
11) $10 - 8i$
12) $6 + 4i$
13) $6 - 4i$
14) $8 + 11i$
15) $8 + 22i$
16) $10 - 4i$

17) $-9 + 18i$
18) $12 + 8i$
19) $4 - 4i$
20) $11 + 7i$
21) $18 - 3i$
22) $22 + 23i$
23) $11 + 18i$
24) $5 + 3i$

25) $-15 + 26i$
26) $-23 + 16i$
27) $-12 + 7i$
28) $-53 + 19i$
29) $33 - 18i$
30) $-11 - 10i$
31) $11 - 51i$
32) $-12 + 9i$

Multiplying and Dividing Complex Numbers

1) 15
2) 16
3) $-27i$
4) 36
5) $-2 - 11i$
6) $21 - 20i$
7) $22 - 32i$
8) $-35 + 12i$

9) $30 - 60i$
10) $14 - 48i$
11) $-23 - 9i$
12) $-20 - 56i$
13) $30 + 24i$
14) $\frac{i}{5}$
15) $\frac{4}{5} + \frac{3}{5}i$

16) $-6 + 3i$
17) $\frac{80}{41} - \frac{100}{41}i$
18) $\frac{9}{4} - \frac{3}{2}i$
19) $-\frac{2}{17} + \frac{8}{17}i$
20) $\frac{19}{20} - \frac{17}{20}i$
21) $-\frac{1}{2} + \frac{5}{2}i$

22) $\frac{25}{17} + \frac{15}{17}i$
23) $\frac{13}{29} - \frac{11}{29}i$
24) $2i$
25) $-\frac{10}{17} + \frac{11}{17}i$

Graphing Complex Numbers

1) $-4 - 3i$
2) $3 + i$
3) $-4 + 3i$
4) $4 - 2i$

Rationalizing Imaginary Denominators

1) $-i$
2) $-\frac{1}{5}i$
3) $\frac{-1}{13}i$
4) $8i$
5) $-\frac{5}{2}i$
6) $4i$

7) $\frac{7}{3}i$
8) $\frac{17}{3}i$
9) $-\frac{4x}{5y}i$
10) $5 + 5i$
11) $11 + 5i$
12) $1 - \frac{21}{4}i$

13) $\frac{32}{17} - \frac{8}{17}i$
14) $\frac{4}{5} - \frac{3}{5}i$
15) $2 + 3i$
16) $-\frac{23}{10} + \frac{1}{10}i$
17) $-\frac{3}{5} - \frac{6}{5}i$
18) $-1 - 2i$

19) $1 - 4i$
20) $-\frac{1}{5} - \frac{2}{5}i$
21) $-\frac{4}{5} - \frac{11}{5}i$
22) $-\frac{44}{29} + \frac{6}{29}i$
23) $\frac{41}{34} + \frac{11}{34}i$
24) $-\frac{3}{29} - \frac{22}{29}i$

Chapter 10 :
Functions Operations and Quadratic

Topics that you'll practice in this chapter:

- ✓ Evaluating Function
- ✓ Adding and Subtracting Functions
- ✓ Multiplying and Dividing Functions
- ✓ Composition of Functions
- ✓ Quadratic Equation
- ✓ Solving Quadratic Equations
- ✓ Quadratic Formula and the Discriminant
- ✓ Quadratic Inequalities
- ✓ Graphing Quadratic Functions
- ✓ Domain and Range of Radical Functions
- ✓ Solving Radical Equations

It's fine to work on any problem, so long as it generates interesting mathematics along the way – even if you don't solve it at the end of the day." – Andrew Wiles

Evaluating Function

Write each of following in function notation.

1) $h = -8x + 3$

2) $k = 2a - 14$

3) $d = 11t$

4) $y = \frac{5}{12}x - \frac{7}{12}$

5) $m = 24n - 210$

6) $c = p^2 - 5p + 10$

Evaluate each function.

7) $f(x) = 2x - 7$, find $f(-3)$

8) $g(x) = \frac{1}{9}x + 12$, find $f(18)$

9) $h(x) = -4x + 9$, find $f(3)$

10) $f(x) = -x + 19$, find $f(-3)$

11) $f(a) = 7a - 12$, find $f(3)$

12) $h(x) = 14 - 3x$, find $f(-4)$

13) $g(n) = 6n - 10$, find $f(2)$

14) $f(x) = -11x - 4$, find $f(-1)$

15) $k(n) = -20 - 3.5n$, find $f(2)$

16) $f(x) = -0.7x + 3.3$, find $f(-7)$

17) $g(n) = \frac{11n+8}{n}$, find $g(2)$

18) $g(n) = \sqrt{3n} + 12$, find $g(3)$

19) $h(x) = x^{-2} - 7$, find $h(\frac{1}{9})$

20) $h(n) = n^{-3} + 11$, find $h(\frac{1}{4})$

21) $h(n) = n^3 - 2$, find $h(\frac{1}{2})$

22) $h(n) = n^2 - 4$, find $h(-\frac{1}{3})$

23) $h(n) = 4n^2 - 13$, find $h(-5)$

24) $h(n) = -2n^3 - 6n$, find $h(2)$

25) $g(n) = \sqrt{16n^2} - \sqrt{n}$, find $g(4)$

26) $h(a) = \frac{-14a+9}{3a}$, find $h(-b)$

27) $k(a) = 12a - 14$, find $k(a - 3)$

28) $h(x) = \frac{1}{9}x + 18$, find $h(-18x)$

29) $h(x) = 8x^2 + 16$, find $h(\frac{x}{2})$

30) $h(x) = x^4 - 20$, find $h(-2x)$

Adding and Subtracting Functions

✎ **Perform the indicated operation.**

1) $f(x) = 2x + 3$

   $g(x) = x + 7$

   Find $(f - g)(2)$

2) $g(a) = -5a - 8$

   $f(a) = -3a - 5$

   Find $(g - f)(-2)$

3) $h(t) = 4t + 3$

   $g(t) = 4t + 7$

   Find $(h - g)(t)$

4) $g(a) = -6a - 10$

   $f(a) = 3a^2 + 9$

   Find $(g - f)(x)$

5) $g(x) = \frac{5}{6}x - 23$

   $h(x) = \frac{5}{12}x + 25$

   Find $g(12) - h(12)$

6) $h(x) = \sqrt{3x} - 2$

   $g(x) = \sqrt{3x} + 5$

   Find $(h + g)(12)$

7) $f(x) = x^{-1}$

   $g(x) = x^2 + \frac{5}{x}$

   Find $(f - g)(-3)$

8) $h(n) = n^2 + 2$

   $g(n) = -4n + 6$

   Find $(h - g)(2a)$

9) $g(x) = -2x^2 - 5 - 4x$

   $f(x) = 7 + 2x$

   Find $(g - f)(3x)$

10) $g(t) = 11t - 4$

    $f(t) = -2t^2 + 5$

    Find $(g + f)(-t)$

11) $f(x) = 8x + 9$

    $g(x) = -5x^2 + 3x$

    Find $(f - g)(-x^2)$

12) $f(x) = -3x^4 - 5x$

    $g(x) = 2x^4 + 5x + 22$

    Find $(f + g)(3x^2)$

Multiplying and Dividing Functions

✎ **Perform the indicated operation.**

1) $g(x) = -2x - 1$

    $f(x) = 4x + 3$

    Find $(g.f)(2)$

2) $f(x) = 5x$

    $h(x) = -2x + 3$

    Find $(f.h)(-2)$

3) $g(a) = 5a - 2$

    $h(a) = 2a - 3$

    Find $(g.h)(-3)$

4) $f(x) = 2x - 7$

    $h(x) = x - 5$

    Find $\left(\frac{f}{h}\right)(4)$

5) $f(x) = 8a^2$

    $g(x) = 3 + 2a$

    Find $\left(\frac{f}{g}\right)(2)$

6) $g(a) = \sqrt{4a} + 2$

    $f(a) = (-a)^4 + 1$

    Find $\left(\frac{g}{f}\right)(1)$

7) $g(t) = t^3 + 1$

    $h(t) = 5t - 2$

    Find $(g.h)(-2)$

8) $g(n) = n^2 + 2n - 4$

    $h(n) = -5n + 3$

    Find $(g.h)(1)$

9) $g(a) = (a - 3)^2$

    $f(a) = a^2 + 4$

    Find $\left(\frac{g}{f}\right)(3)$

10) $g(x) = -3x^2 + \frac{4}{5}x + 9$

    $f(x) = x^2 - 24$

    Find $\left(\frac{g}{f}\right)(5)$

11) $f(x) = 2x^3 - 5x^2 + 1$

    $g(x) = 3x - 1$

    Find $(f.g)(x)$

12) $f(x) = 5x - 2$

    $g(x) = x^3 - 2x$

    Find $(f.g)(x^2)$

Composition of Functions

✎Using $f(x) = 2x - 5$ and $g(x) = -2x$, find:

1) $f(g(2)) =$

4) $g(f(5)) =$

2) $f(g(-1)) =$

5) $f(g(3)) =$

3) $g(f(-4)) =$

6) $g(f(0)) =$

✎Using $f(x) = -\frac{1}{4}x + \frac{3}{4}$ and $g(x) = 2x^2$, find:

7) $g(f(-2)) =$

10) $f(f(1)) =$

8) $g(f(4)) =$

11) $g(f(-4)) =$

9) $g(g(1)) =$

12) $f(g(x)) =$

✎Using $f(x) = -2x + 2$ and $g(x) = x + 1$, find:

13) $g(f(1)) =$

16) $f(g(-3)) =$

14) $f(f(0)) =$

17) $g(f(2)) =$

15) $f(g(-1)) =$

18) $f(g(x)) =$

✎Using $f(x) = \sqrt{x + 9}$ and $g(x) = x - 9$, find:

19) $f(g(9)) =$

22) $f(f(7)) =$

20) $g(f(-9)) =$

23) $g(f(-5)) =$

21) $f(g(4)) =$

24) $g(g(0)) =$

Quadratic Equation

✎ Multiply.

1) $(x - 4)(x + 6) =$ _____

2) $(x + 5)(x + 7) =$ _____

3) $(x - 6)(x + 8) =$ _____

4) $(x + 2)(x - 9) =$ _____

5) $(x - 7)(x - 8) =$ _____

6) $(3x + 2)(x - 3) =$ _____

7) $(4x - 3)(x + 2) =$ _____

8) $(4x - 5)(x + 1) =$ _____

9) $(7x + 1)(x - 6) =$ _____

10) $(5x + 1)(3x - 3) =$ _____

✎ Factor each expression.

11) $x^2 - 2x - 8 =$ _____

12) $x^2 + 8x + 15 =$ _____

13) $x^2 - 2x - 24 =$ _____

14) $x^2 - 10x + 21 =$ _____

15) $x^2 + 10x + 21 =$ _____

16) $4x^2 + 9x + 5 =$ _____

17) $5x^2 + 13x - 6 =$ _____

18) $5x^2 + 17x - 12 =$ _____

19) $2x^2 + 7x + 5 =$ _____

20) $9x^2 - 21x + 6 =$ _____

✎ Calculate each equation.

21) $(x + 6)(x - 3) = 0$

22) $(x + 1)(x + 8) = 0$

23) $(3x + 6)(x + 5) = 0$

24) $(2x - 2)(4x + 8) = 0$

25) $x^2 + x + 10 = 22$

26) $x^2 + 11x + 36 = 12$

27) $2x^2 + 9x + 9 = 5$

28) $x^2 + 3x - 24 = 4$

29) $5x^2 + 5x - 40 = 20$

30) $8x^2 + 8x = 48$

Solving Quadratic Equations

✎ **Solve each equation by factoring or using the quadratic formula.**

1) $(x + 9)(x - 1) = 0$

2) $(x + 7)(x + 6) = 0$

3) $(x - 8)(x + 3) = 0$

4) $(x - 6)(x - 4) = 0$

5) $(x + 2)(x + 12) = 0$

6) $(5x + 4)(x + 7) = 0$

7) $(6x + 1)(4x + 5) = 0$

8) $(2x + 7)(x + 8) = 0$

9) $(x + 6)(3x + 15) = 0$

10) $(12x + 2)(x + 8) = 0$

11) $x^2 = 8x$

12) $x^2 - 16 = 0$

13) $3x^2 + 6 = 9x$

14) $-2x^2 - 8 = 10x$

15) $5x^2 + 40x = 45$

16) $x^2 + 10x = 24$

17) $x^2 + 6x = 16$

18) $x^2 + 9x = -18$

19) $x^2 + 13x = -36$

20) $x^2 + 3x - 15 = 5x$

21) $x^2 + 8x + 7 = -8$

22) $3x^2 - 11x = -9 + x$

23) $10x^2 + 3 = 27x - 15$

24) $7x^2 - 6x + 8 = 8$

25) $2x^2 - 12 = -3x + 2$

26) $10x^2 - 26x - 3 = -15$

27) $3x^2 + 21 = -16x + 5$

28) $x^2 + 15x - 10 = -66$

29) $3x^2 - 8x - 8 = 4 + x$

30) $2x^2 + 6x - 24 = 12$

31) $3x^2 - 33x + 54 = -18$

32) $-10x^2 - 15x - 9 = -9 - 27x^2$

Quadratic Formula and the Discriminant

✎ Find the value of the discriminant of each quadratic equation.

1) $3x(x - 8) = 0$

2) $2x^2 + 6x - 4 = 0$

3) $x^2 + 6x + 7 = 0$

4) $x^2 - x + 3 = 0$

5) $x^2 + 4x - 3 = 0$

6) $2x^2 + 6x - 10 = 0$

7) $3x^2 + 7x + 5 = 0$

8) $x^2 - 6x - 4 = 0$

9) $2x^2 + 8x + 3 = 0$

10) $x^2 + 7x - 5 = 0$

11) $5x^2 + 2x - 3 = 0$

12) $-3x^2 - 11x + 4 = 0$

13) $-6x^2 - 12x + 8 = 0$

14) $-x^2 - 9x - 12 = 0$

15) $7x^2 - 6x - 10 = 0$

16) $-4x^2 - 2x + 8 = 0$

17) $5x^2 + 8x - 2 = 0$

18) $6x^2 - 4x = 0$

19) $3x^2 - 5x + 2 = 0$

20) $4x^2 + 9x + 3 = 0$

✎ Find the discriminant of each quadratic equation then state the number of real and imaginary solutions.

21) $-4x^2 - 16 = 16x$

22) $20x^2 = 20x - 5$

23) $-11x^2 - 19x = 26$

24) $22x^2 - 4x + 1 = 18x^2$

25) $-11x^2 = -15x + 8$

26) $3x^2 + 6x + 9 = 6$

27) $13x^2 - 5x - 12 = -26$

28) $-8x^2 - 32x - 25 = 7$

Graphing Quadratic Functions

✎ketch the graph of each function. Identify the vertex and axis of symmetry.

1) $y = (x + 3)^2 + 2$

2) $y = (x - 3)^2 - 2$

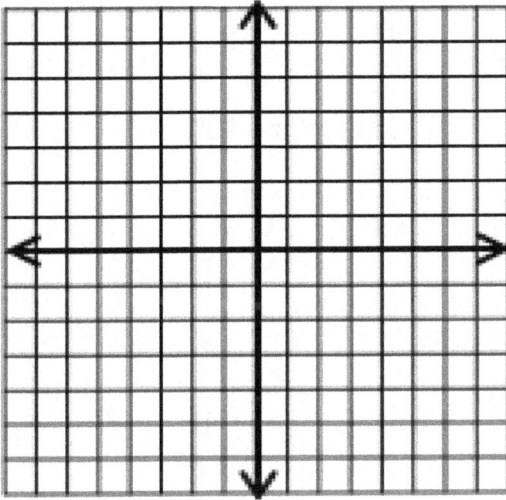
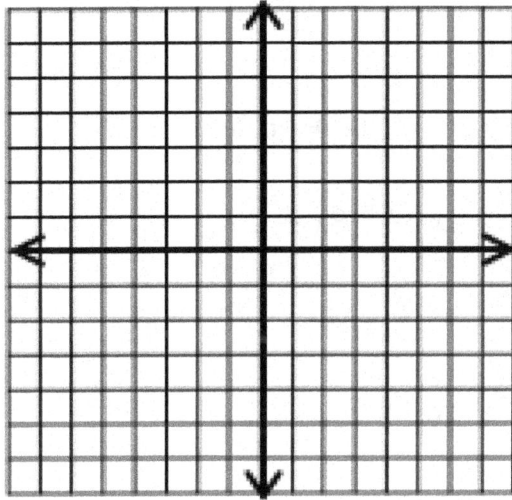

3) $y = 6 - (-x + 4)^2$

4) $y = -3x^2 - 6x + 9$

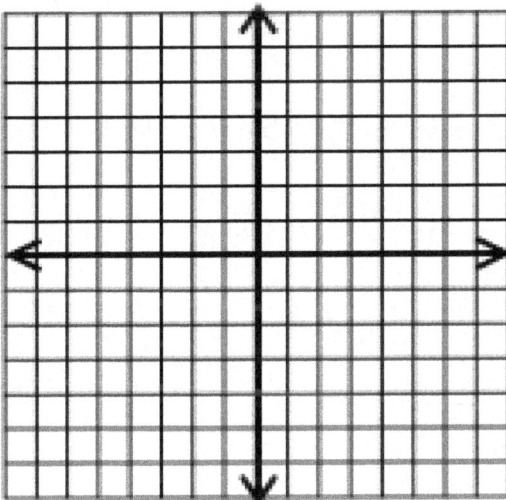
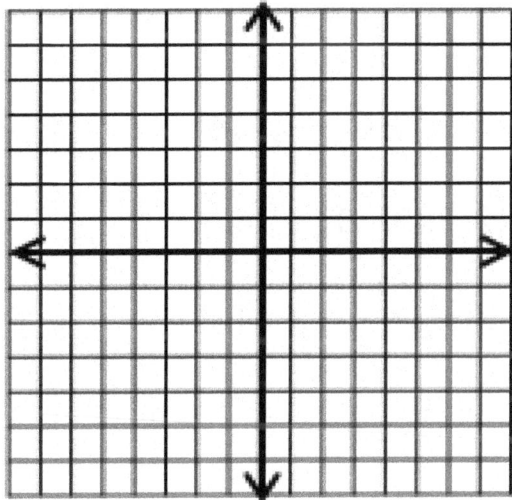

Quadratic Inequalities

✍ **Solve each quadratic inequality.**

1) $x^2 - 25 < 0$

2) $-x^2 - 6x - 8 > 0$

3) $5x^2 + 15x + 30 < 0$

4) $x^2 + 8x + 16 > 0$

5) $2x^2 - 18x - 20 \geq 0$

6) $x^2 > -10x - 25$

7) $3x^2 + 2x + 16 \leq 0$

8) $x^2 - 5x - 14 \leq 0$

9) $x^2 - 6x - 7 \geq 0$

10) $2x^2 + 16x - 18 < 0$

11) $x^2 + 6x - 72 > 0$

12) $3x^2 - 3x - 36 > 0$

13) $x^2 - 15x + 64 \leq 0$

14) $2x^2 - 24x + 72 \leq 0$

15) $x^2 - 16x + 63 \geq 0$

16) $x^2 - 16x + 55 \geq 0$

17) $x^2 - 81 \leq 0$

18) $x^2 - 17x + 42 \geq 0$

19) $9x^2 + 14x + 36 \leq 0$

20) $4x^2 - 2x - 24 > 2x^2$

21) $5x^2 - 20x + 20 < 0$

22) $7x^2 - 6x \geq 6x^2 - 5$

23) $5x^2 - 15 > 4x^2 + 2x$

24) $3x^2 - 4x \geq 3x^2 - 9x + 15$

25) $8x^2 + 9x - 54 > 5x^2$

26) $10x^2 + 50x - 60 < 0$

27) $-x^2 + 15x - 57 \geq 0$

28) $-5x^2 + 25x + 30 \leq 0$

29) $5x^2 + 40x + 75 < 0$

30) $9x^2 + 20x + 180 \leq 0$

31) $3x^2 + 2x - 36 \geq -x$

32) $3x^2 + 9x + 9 \leq 6x^2 + 3x$

Domain and Range of Radical Functions

✍ **Identify the domain and range of each function.**

1) $y = \sqrt{x + 8} - 7$

2) $y = \sqrt[3]{3x - 5} - 4$

3) $y = \sqrt{3x - 9} + 3$

4) $y = \sqrt[3]{(4x + 6)} - 2$

5) $y = 3\sqrt{4x + 20} + 6$

6) $y = \sqrt[3]{(5x - 2)} - 11$

7) $y = 4\sqrt{9x^2 + 8} + 3$

8) $y = \sqrt[3]{(7x^2 - 2)} - 6$

9) $y = 2\sqrt{2x^3 + 16} - 3$

10) $y = \sqrt[3]{(11x + 4)} - 2x$

11) $y = 3\sqrt{-2(4x + 8)} + 5$

12) $y = \sqrt[5]{(3x^2 - 12)} - 6$

13) $y = 3\sqrt{x - 5} - 2$

14) $y = \sqrt[3]{6x + 9} - 4$

✍ **Sketch the graph of each function.**

15) $y = -3\sqrt{x} + 5$

16) $y = 3\sqrt{x} - 6$

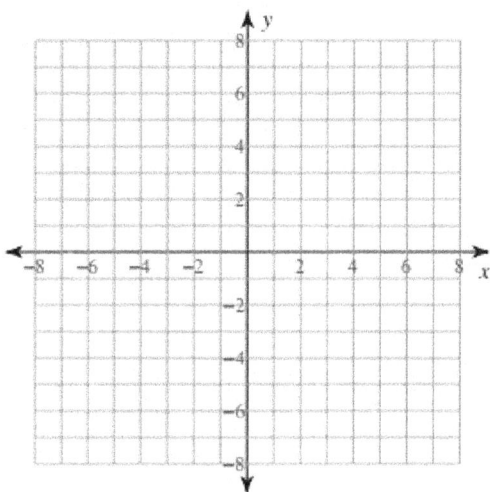

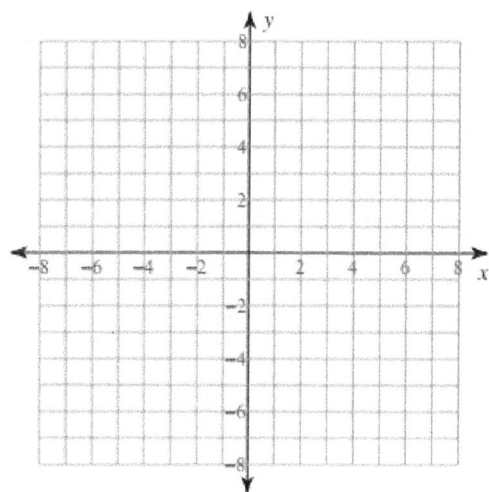

Solving Radical Equations

✎ **Solve each equation. Remember to check for extraneous solutions.**

1) $\sqrt{a} = 9$

2) $\sqrt{v} = 6$

3) $\sqrt{r} = 4$

4) $8 = 16\sqrt{x}$

5) $\sqrt{x+3} = 18$

6) $6 = \sqrt{x-7}$

7) $4 = \sqrt{r-3}$

8) $\sqrt{x-5} = 7$

9) $12 = \sqrt{x-4}$

10) $\sqrt{m+5} = 8$

11) $7\sqrt{5a} = 35$

12) $6\sqrt{2x} = 48$

13) $2 = \sqrt{6x-32}$

14) $\sqrt{304-4x} = 4$

15) $\sqrt{r+2} - 8 = 6$

16) $-21 = -7\sqrt{r+9}$

17) $60 = 6\sqrt{5v}$

18) $x = \sqrt{40-3x}$

19) $\sqrt{90-27a} = 3a$

20) $\sqrt{-8n+88} = 4$

21) $\sqrt{15r-5} = 4r-3$

22) $\sqrt{-64+32x} = 4x$

23) $\sqrt{4x+15} = \sqrt{2x+11}$

24) $\sqrt{12v} = \sqrt{15v-21}$

25) $\sqrt{9-x} = \sqrt{x-3}$

26) $\sqrt{6m+34} = \sqrt{8m+34}$

27) $\sqrt{7r+32} = \sqrt{-8-3r}$

28) $\sqrt{4k+10} = \sqrt{2-4k}$

29) $-20\sqrt{x-13} = -40$

30) $\sqrt{90-2x} = \sqrt{\dfrac{x}{4}}$

Answers of Worksheets

Evaluating Function

1) $h(x) = -8x + 3$
2) $k(a) = 2a - 14$
3) $d(t) = 11t$
4) $f(x) = \frac{5}{12}x - \frac{7}{12}$
5) $m(n) = 24n - 210$
6) $c(p) = p^2 - 5p + 10$
7) -13
8) 14
9) -3
10) 22

11) 9
12) 26
13) 2
14) 7
15) -27
16) 8.2
17) 15
18) 15
19) 74
20) 75

21) $-1\frac{7}{8}$
22) $-3\frac{8}{9}$
23) 87
24) -28
25) 14
26) $-\frac{14b+9}{3b}$
27) $12a - 50$
28) $-2x + 18$
29) $2x^2 + 16$
30) $16x^4 - 20$

Adding and Subtracting Functions

1) -2
2) 1
3) -4
4) $-3x^2 - 6x - 19$

5) -43
6) 15
7) $-7\frac{2}{3}$
8) $4a^2 + 8a - 4$

9) $-18x^2 - 18x - 12$
10) $-2t^2 - 11t + 1$
11) $5x^4 - 5x^2 + 9$
12) $-81x^8 + 22$

Multiplying and Dividing Functions

1) -55
2) -70
3) 153
4) -1

5) $4\frac{4}{7}$
6) 2
7) 84

8) 2
9) 0
10) -62

11) $6x^4 - 17x^3 + 5x^2 + 3x - 1$

12) $5x^8 - 2x^6 - 10x^4 + 4x^2$

Composition of Functions

1) -13
2) -1
3) 26
4) -10
5) -17

6) 10
7) $\frac{25}{8}$
8) $\frac{1}{8}$
9) 8

10) $\frac{5}{8}$
11) $\frac{49}{8}$
12) $-\frac{1}{2}(x^2 - \frac{3}{2})$
13) 1

14) -2
15) 2
16) 6
17) -1
18) $-2x$

19) 3   21) 2   23) −7

20) −9   22) $\sqrt{13}$ 24) −18

Quadratic Equations

1) $x^2 + 2x - 24$

2) $x^2 + 12x + 35$

3) $x^2 + 2x - 48$

4) $x^2 - 7x - 18$

5) $x^2 - 15x + 56$

6) $3x^2 - 7x - 6$

7) $4x^2 + 5x - 6$

8) $4x^2 - x - 5$

9) $7x^2 - 41x - 6$

10) $15x^2 - 12x - 3$

11) $(x - 4)(x + 2)$

12) $(x + 5)(x + 3)$

13) $(x - 6)(x + 4)$

14) $(x - 3)(x - 7)$

15) $(x + 3)(x + 7)$

16) $(4x + 5)(x + 1)$

17) $(5x - 2)(x + 3)$

18) $(5x - 3)(x + 4)$

19) $(2x + 5)(x + 1)$

20) $3(x - 2)(3x - 1)$

21) $x = -6, x = 3$

22) $x = -1, x = -8$

23) $x = -2, x = -5$

24) $x = 1, x = -2$

25) $x = 3, x = -4$

26) $x = -3, x = -8$

27) $x = -4, x = -\frac{1}{2}$

28) $x = 4, x = -7$

29) $x = 3, x = -4$

30) $x = -3, x = 2$

Solving quadratic equations

1) $\{-9, 1\}$

2) $\{-6, -7\}$

3) $\{8, -3\}$

4) $\{6, 4\}$

5) $\{-2, -12\}$

6) $\{-\frac{4}{5}, -7\}$

7) $\{-\frac{5}{4}, -\frac{1}{6}\}$

8) $\{-\frac{7}{2}, -8\}$

9) $\{-6, -5\}$

10) $\{-\frac{1}{6}, -8\}$

11) $\{8, 0\}$

12) $\{4, -4\}$

13) $\{2, 1\}$

14) $\{-4, -1\}$

15) $\{1, -9\}$

16) $\{2, -12\}$

17) $\{2, -8\}$

18) $\{-3, -6\}$

19) $\{-4, -9\}$

20) $\{5, -3\}$

21) $\{-5, -3\}$

22) $\{1, 3\}$

23) $\{\frac{6}{5}, \frac{3}{2}\}$

24) $\{\frac{6}{7}, 0\}$

25) $\{-\frac{7}{2}, 2\}$

26) $\{\frac{3}{5}, 2\}$

27) $\{-\frac{4}{3}, -4\}$

28) $\{-8, -7\}$

29) $\{4, -1\}$

30) $\{3, -6\}$

31) $\{3, 8\}$

32) $\{\frac{15}{17}, 0\}$

Quadratic formula and the discriminant

1) 576

2) 68

3) 8

4) −11

5) 28

6) 116

7) −11

8) 52

9) 40

10) 69

11) 64

12) 169

13) 336

14) 33

15) 316

16) 132

17) 104

18) 16

19) 1

20) 33

21) 0, *one real solution*

22) 0, *one real solution*

23) −783, *no solution*

24) $0,$ *one real solution*   26) $0,$ *one real solution*   28) $0,$ *one real solution*

25) $-127,$ *no solution*   27) $-703,$ *no solution*

Graphing quadratic functions

1) $(-3, 2), x = -3$

2) $(3, -2), x = 3$

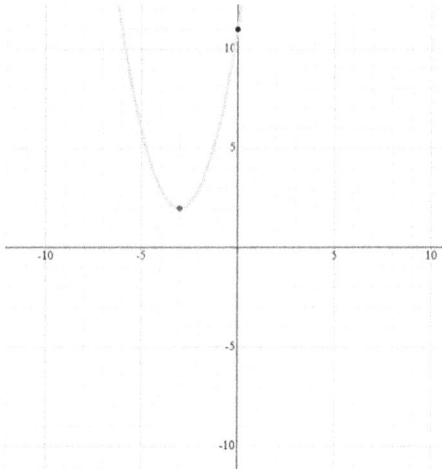

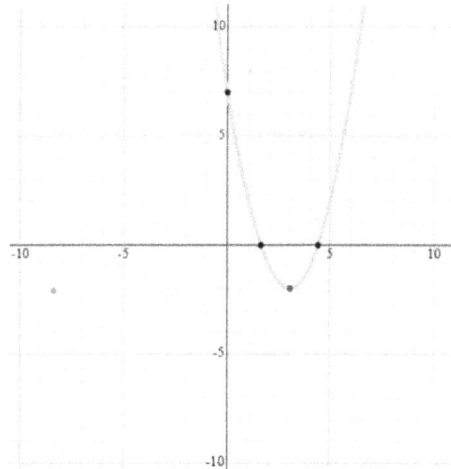

3) $(4, 6), x = 4$

4) $(-1, 12), x = -1$

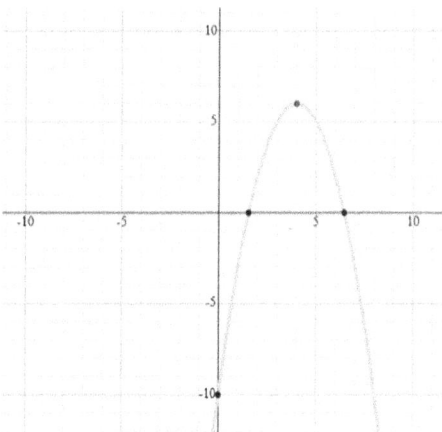

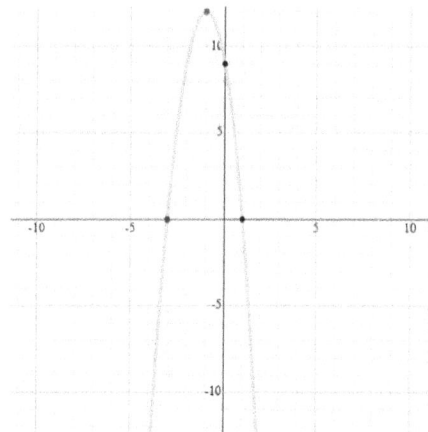

Quadratic inequalities

1) $-5 < x < 5$

7) no solution

13) no solution

2) $-4 < x < -2$

8) $-2 \leq x \leq 7$

14) $x = 6$

3) no solution

9) $x \leq -1 \ or \ x \geq 7$

15) $x \leq 7 or \ x \geq 9$

4) $x < -4 \ or \ x > -4$

10) $-9 < x < 1$

16) $x \leq 5 or \ x \geq 11$

5) $x \leq -1 \ or \ x \geq 10$

11) $x < -12 \ or \ x > 6$

17) $-9 \leq x \leq 9$

6) $x < -5 \ or \ x > -5$

12) $-3 < x < 4$

18) $x \leq 3 \ or \ x \geq 14$

19) no solution

20) $x < -3$ or $x > 4$

21) no solution

22) $x \leq 1$ or $x \geq 5$

23) $x < -3$ or $x > 5$

24) $x \geq 3$

25) $x < -6$ or $x > 3$

26) $-6 < x < 1$

27) no solution

28) $x \leq -1$ or $x \geq 6$

29) $-5 < x < -3$

30) no solution

31) $x \leq -4$ or $x \geq 3$

32) $x \leq -1$ or $x \geq 3$

Domain and range of radical functions

1) domain: $x \geq -8$

   range: $y \geq -7$

2) domain: {all real numbers}

   range: {all real numbers}

3) domain: $x \geq 3$

   range: $y \geq 3$

4) domain: {all real numbers}

   range: {all real numbers}

5) domain: $x \geq -5$

   range: $y \geq 6$

6) domain: {all real numbers}

   range: {all real numbers}

7) domain: {all real numbers}

   range: $y \geq 8\sqrt{2} + 3$

8) domain: {all real numbers}

   range: {all real numbers}

9) domain: $x \geq -2$

   range: $y \geq -3$

10) domain: {all real numbers}

    range: {all real numbers}

11) domain: $x \leq -2$

    range: $y \geq 5$

12) domain: {all real numbers}

    range: {all real numbers}

13) domain: $x \geq 5$

    range: $y \geq -2$

14) domain: {all real numbers}

    range: {all real numbers}

15)

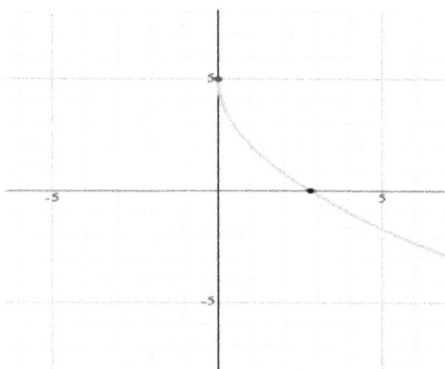

16)

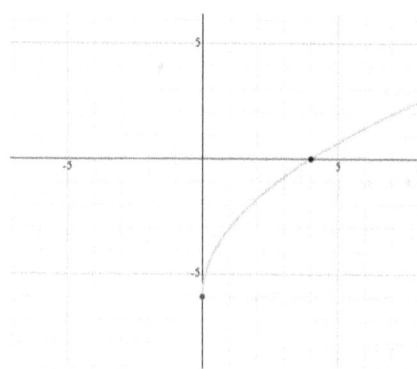

Solving radical equations

1) {81}

2) {36}

3) {16}

4) {$\frac{1}{4}$}

5) {321}

6) {43}

7) {19}

8) {54}

9) {148}

10) {59}

11) {5}

12) {32}

13) {6}

14) {72}

15) {194}

16) {0}

17) {20}

18) {5}

19) {2}

20) {9}

21) {2}

22) no solution

23) {−2}

24) {7}

25) {6}

26) {0}

27) {−4}

28) {−1}

29) {17}

30) {40}

Chapter 11 :

Sequences and Series

Topics that you'll practice in this chapter:

- ✓ Arithmetic Sequences
- ✓ Geometric Sequences
- ✓ Comparing Arithmetic and Geometric Sequences
- ✓ Finite Geometric Series
- ✓ Infinite Geometric Series

"The fact that people will be full of greed, fear, or folly is predictable. The sequence is not predictable." -Warren Buffett

Arithmetic Sequences

✎ **Find the next three terms of each arithmetic sequence.**

1) $32, 26, 20, 14, 8, \ldots$

2) $-56, -44, -32, -20, \ldots$

3) $17, 26, 35, 44, 53, \ldots$

4) $5, 11, 17, 23, 29, \ldots$

✎ **Given the first term and the common difference of an arithmetic sequence find the first five terms and the explicit formula.**

5) $a_1 = 20, d = 3$

6) $a_1 = -11, d = -5$

7) $a_1 = 32, d = 6$

8) $a_1 = 240, d = -80$

✎ **Given a term in an arithmetic sequence and the common difference find the first five terms and the explicit formula.**

9) $a_{20} = -500, d = -50$

10) $a_{24} = 98, d = 7$

11) $a_{51} = -88.2, d = -5.2$

12) $a_{68} = -980, d = -27$

✎ **Given a term in an arithmetic sequence and the common difference find the recursive formula and the three terms in the sequence after the last one given.**

13) $a_{21} = -187, d = -9$

14) $a_{12} = 63.5, d = 5.2$

15) $a_{31} = 58.2, d = 1.8$

16) $a_{42} = 6.8, d = 0.4$

Geometric Sequences

✎ **Determine if the sequence is geometric. If it is, find the common ratio.**

1) $1, -7, 49, -343, \ldots$

3) $8, 24, 48, 240, \ldots$

2) $-3, -12, -48, -192, \ldots$

4) $-5, -10, -20, -40, \ldots$

✎ **Given the first term and the common ratio of a geometric sequence find the first five terms and the explicit formula.**

5) $a_1 = 0.4, r = -3$

6) $a_1 = 0.2, r = 4$

✎ **Given the recursive formula for a geometric sequence find the common ratio, the first five terms, and the explicit formula.**

7) $a_n = a_{n-1} \times 4, a_1 = 2$

9) $a_n = a_{n-1} \cdot 5, a_1 = 0.2$

8) $a_n = a_{n-1} \cdot (-2), a_1 = -4$

10) $a_n = a_{n-1} \cdot 3, a_1 = -3$

✎ **Given two terms in a geometric sequence find the 6th term and the recursive formula.**

11) $a_3 = 576$ and $a_5 = 36$

12) $a_2 = -0.4$ and $a_4 = -1.6$

Comparing Arithmetic and Geometric Sequences

✎ **For each sequence, state if it is arithmetic, geometric, or neither.**

1) $6, 11, 16, 21, \dots$

2) $2, 5, 8, 11, \dots$

3) $3, 10, 20, 27, \dots$

4) $1, 11, 22, 33, 44, \dots$

5) $4, 8, 10, 24, 96, \dots$

6) $2, 10, 20, 50, 200, \dots$

7) $0.2, 1, 5, 25, 125, \dots$

8) $4, 12, 36, 108, \dots$

9) $-27, -34, -41, -48, -55, \dots$

10) $-3, 15, -75, 375, -1875, \dots$

11) $10, 25, 50, 65, 80, \dots$

12) $5, 15, 150, 250, 350 \dots$

13) $-35, -20, -5, 10, 25, \dots$

14) $a_n = 4 \cdot 8^{n-1}$

15) $a_n = 3 \cdot 6^{n-1}$

16) $a_n = 8 - 4n$

17) $a_n = -210 + 310n$

18) $a_n = 53 + 53n$

19) $a_n = -10 \cdot (-5)^{n-1}$

20) $a_n = 49 + 63n$

21) $a_n = (3n)^4$

22) $a_n = 40 + 8n$

23) $a_n = -(13)^{n-1}$

24) $a_n = -10 \cdot (0.2)^{n-1}$

25) $a_n = \frac{3n+2}{4^n}$

26) $a_n = \frac{47+13n}{7^n}$

27) $a_n = \frac{12-12n}{12^n}$

28) $a_n = \frac{32-a_{n-1}}{2n}$

29) $a_n = \frac{4}{15} - \frac{2}{7}n$

Finite Geometric Series

📌 **Evaluate the related series of each sequence.**

1) $-2, 4, -8, 16$

2) $-1, 6, -36, 216, -1,296$

3) $-2, 8, -32, 128, -512$

4) $4, 8, 16, 32, 64$

5) $-4, -12, -36, -108$

6) $5, -15, 45, -135, 405$

📌 **Evaluate each geometric series described.**

7) $1 + 5 + 25 + 125 \dots, n = 5$ _____

8) $1 - 6 + 36 - 216 \dots, n = 6$ _____

9) $-2 - 6 - 18 - 54 \dots, n = 8$ _____

10) $0.2 - 1 + 5 - 25 \dots, n = 6$ _____

11) $0.6 - 3.6 + 21.6 - 129.6 \dots, n = 5$ _____

12) $-1 - 4 - 16 - 64 \dots, n = 6$ _____

13) $a_1 = -2, r = 6, n = 5$ _____

14) $a_1 = 1, r = 9, n = 6$ _____

15) $\sum_{n=1}^{6} 4 \cdot (-3)^{n-1}$ _____

16) $\sum_{n=1}^{7} 2 \cdot (-5)^{n-1}$ _____

17) $\sum_{n=1}^{10} 0.1 \cdot (2)^{n-1}$ _____

18) $\sum_{m=1}^{6} (-4)^{m-1}$ _____

19) $\sum_{m=1}^{6} 5 \times (3)^{m-1}$ _____

20) $\sum_{k=1}^{5} 7 \times (6)^{k-1}$ _____

Infinite Geometric Series

👈 **Determine if each geometric series converges or diverges.**

1) $a_1 = -1.8, \ r = 6$

2) $a_1 = 10.8, r = 0.3$

3) $a_1 = -2, r = 6.1$

4) $a_1 = 5, r = 0.24$

5) $a_1 = 1.2, r = 8$

6) $-1, 6, -36, 216, \ldots$

7) $6, -1, \dfrac{1}{6}, -\dfrac{1}{36}, \dfrac{1}{216}, \ \ldots$

8) $512 + 64 + 8 + 1 \ \ldots$

9) $-5 + \dfrac{15}{7} - \dfrac{45}{49} + \dfrac{135}{343} \ \ldots$

10) $\dfrac{400}{459} - \dfrac{200}{153} + \dfrac{100}{51} - \dfrac{50}{17} \ \ldots$

👈 **Evaluate each infinite geometric series described.**

11) $a_1 = 2, r = -\dfrac{1}{4}$

12) $a_1 = 36, r = -\dfrac{1}{6}$

13) $a_1 = 9, r = \dfrac{1}{3}$

14) $a_1 = 12, r = \dfrac{1}{7}$

15) $1 + 0.2 + 0.04 + 0.008 + \cdots$

16) $64 - 16 + 4 - 1 \ \ldots,$

17) $1 - 0.3 + 0.09 - 0.027 \ \ldots,$

18) $-5 + \dfrac{15}{7} - \dfrac{45}{49} + \dfrac{135}{343} \ \ldots,$

19) $\sum_{k=1}^{\infty} 7^{k-1}$

20) $\sum_{i=1}^{\infty} \left(\dfrac{2}{5}\right)^{i-1}$

21) $\sum_{k=1}^{\infty} \left(-\dfrac{2}{9}\right)^{k-1}$

22) $\sum_{n=1}^{\infty} 6\left(\dfrac{1}{3}\right)^{n-1}$

Answers of Worksheets

Arithmetic Sequences

1) $2, -4, -10$

2) $-8, 4, 16$

3) $62, 71, 80$

4) $35, 41, 47$

5) First Five Terms: $20, 23, 26, 29, 32$, Explicit: $a_n = 20 + 3(n - 1)$

6) First Five Terms: $-11, -16, -21, -26, -31$, Explicit: $a_n = -11 - 5(n - 1)$

7) First Five Terms: $32, 38, 44, 50, 56$, Explicit: $a_n = 32 + 6(n - 1)$

8) First Five Terms: $240, 160, 80, 0, -80$, Explicit: $a_n = 240 - 80(n - 1)$

9) First Five Terms: $450, 400, 350, 300, 250$, Explicit: $a_n = 450 - 50(n - 1)$

10) First Five Terms: $-63, -56, -49, -42, -35$, Explicit: $a_n = -63 + 7(n - 1)$

11) First Five Terms: $171.8, 166.6, 161.4, 156.2, 151$, Explicit: $a_n = 171.8 - 5.2(n - 1)$

12) First Five Terms: $829, 802, 775, 748, 721$, Explicit: $a_n = 829 - 27(n - 1)$

13) Next 3 terms: $-196, -205, -214$, Recursive: $a_n = a_{n-1} - 9, a_1 = -7$

14) Next 3 terms: $68.7, 73.9, 79.1, 84.3$ Recursive: $a_n = a_{n-1} + 5.2, \ a_1 = 6.3$

15) Next 3 terms: $60, 61.8, 63.6$, Recursive: $a_n = a_{n-1} + 1.8, a_1 = 4.2$

16) Next 3 terms: $7.2, 7.6, 8$, Recursive: $a_n = a_{n-1} + 0.4, a_1 = -9.6$

Geometric Sequences

1) $r = -7$

2) $r = 4$

3) not geometric

4) $r = 2$

5) First Five Terms: $0.4, -1.2, 3.6, -10.8, 32.4$

   Explicit: $a_n = 0.4 \times (-3)^{n-1}$

6) First Five Terms: $0.2, 0.8, 3.2, 12.8, 51.2$

   Explicit: $a_n = 0.2 \times (4)^{n-1}$

7) Common Ratio: $r = 4$

   First Five Terms: $2, 8, 32, 128, 512$

   Explicit: $a_n = 2 . (4)^{n-1}$

8) Common Ratio: $r = -2$

First Five Terms: $-4, 8, -16, 32, -64$

Explicit: $a_n = -4 \cdot (-2)^{n-1}$

9) Common Ratio: $r = 5$

First Five Terms: $0.2, 1, 5, 25, 125, 625$

Explicit: $a_n = 0.2 \cdot (5)^{n-1}$

10) Common Ratio: $r = 3$

First Five Terms: $-3, -9, -27, -81, -243$

Explicit: $a_n = -3 \cdot (3)^{n-1}$

11) $a_6 = -9$, Recursive: $a_n = a_{n-1} \cdot (\frac{-1}{4})$, $a_1 = 9,216$

12) $a_6 = -6.4$, Recursive: $a_n = a_{n-1} \cdot (-2)$, $a_1 = 0.2$

Comparing Arithmetic and Geometric Sequences

1) Arithmetic	11) Neither	21) Neither
2) Arithmetic	12) Neither	22) Arithmetic
3) Neither	13) Arithmetic	23) Geometric
4) Neither	14) Geometric	24) Geometric
5) Neither	15) Geometric	25) Neither
6) Neither	16) Arithmetic	26) Neither
7) Geometric	17) Arithmetic	27) Neither
8) Geometric	18) Arithmetic	28) Neither
9) Arithmetic	19) Geometric	29) Arithmetic
10) Geometric	20) Arithmetic	

Finite Geometric

1) 10	8) $-6,665$	15) -728
2) $-1,111$	9) $-6,560$	16) 26,042
3) -410	10) -520.8	17) 102.3
4) 124	11) 666.6	18) -819
5) -160	12) $-1,365$	19) 1,820
6) 305	13) $-3,110$	20) 10,885
7) 781	14) 66,430	

Infinite Geometric

1) Diverges

2) Converges

3) Diverges

4) Converges

5) Diverges

6) Diverges

7) Converges

8) Converges

9) Converges

10) Converges

11) $\frac{8}{5}$

12) $\frac{216}{7}$

13) $\frac{27}{2}$

14) 14

15) $\frac{5}{4}$

16) $\frac{256}{5}$

17) $\frac{3}{4}$

18) $-\frac{7}{2}$

19) Infinite

20) $\frac{5}{3}$

21) $\frac{9}{11}$

22) 9

Chapter 12 :

Logarithms

Topics that you'll practice in this chapter:

✓ Rewriting Logarithms

✓ Evaluating Logarithms

✓ Properties of Logarithms

✓ Natural Logarithms

✓ Exponential Equations Requiring Logarithms

✓ Solving Logarithmic Equations

Mathematics is an art of human understanding.

— William Thurston

Rewriting Logarithms

✍ **Rewrite each equation in exponential form.**

1) $\log_3 27 = 3$

2) $\log_2 128 = 7$

3) $\log_6 1{,}296 = 4$

4) $\log_5 625 = 4$

5) $\log_{11} 121 = 2$

6) $\log_{12} 1{,}728 = 3$

7) $\log_9 729 = 3$

8) $\log_3 729 = 6$

9) $\log_{10} 10{,}000 = 4$

10) $\log_7 343 = 3$

11) $\log_4 1{,}024 = 5$

12) $\log_{12} 144 = 2$

13) $\log_{13} 2{,}197 = 3$

14) $\log_{25} 5 = \frac{1}{2}$

15) $\log_{81} 3 = \frac{1}{4}$

16) $\log_{3{,}125} 5 = \frac{1}{5}$

17) $\log_{1{,}000} 10 = \frac{1}{3}$

18) $\log_5 \frac{1}{125} = -3$

19) $\log_4 \frac{1}{16} = -2$

20) $\log_a \frac{7}{4} = b$

✍ **Rewrite each exponential equation in logarithmic form.**

21) $2^5 = 32$

22) $4^3 = 64$

23) $5^4 = 625$

24) $11^3 = 1{,}331$

25) $3^5 = 243$

26) $6^4 = 1{,}296$

27) $7^4 = 2{,}401$

28) $9^3 = 729$

29) $4^{-5} = \frac{1}{1{,}024}$

30) $3^{-8} = \frac{1}{6{,}561}$

31) $11^{-2} = \frac{1}{121}$

32) $12^{-3} = \frac{1}{1{,}728}$

33) $4^{-5} = \frac{1}{1{,}024}$

34) $10^{-5} = \frac{1}{100{,}000}$

Evaluating Logarithms

✍ **Evaluate each logarithm.**

1) $\log_3 729 =$

2) $\log_2 256 =$

3) $\log_3 243 =$

4) $\log_4 64 =$

5) $\log_8 64 =$

6) $\log_{11} 121 =$

7) $\log_{10} 10,000 =$

8) $\log_5 \frac{1}{25} =$

9) $\log_4 \frac{1}{256} =$

10) $\log_2 \frac{1}{32} =$

11) $\log_6 \frac{1}{36} =$

12) $\log_9 \frac{1}{81} =$

13) $\log_{12} \frac{1}{144} =$

14) $\log_{1,000} \frac{1}{10} =$

15) $\log_{243} 3 =$

16) $\log_4 \frac{1}{16} =$

17) $\log_8 \frac{1}{512} =$

18) $\log_3 \frac{1}{81} =$

✍ **Circle the points which are on the graph of the given logarithmic**

functions.

19) $y = 4\log_4(3x - 2) + 1$ $\quad (3,4), \quad (2,5), \quad (7,4)$

20) $y = 5\log_6(12x) - 7$ $\quad (2,-2), \quad (\frac{1}{3},12), \quad (\frac{1}{2},-2)$

21) $y = -2\log_3 9(x - 5) + 5$ $\quad (6,-3), \quad (8,-1), \quad (1,6)$

22) $y = \frac{1}{4}\log_6(6x) + \frac{1}{2}$ $\quad (6,1), \quad (6,\frac{1}{4}), \quad (4,\frac{1}{4})$

23) $y = -2\log_8 8(x + 4) + 9$ $\quad (-4,0), \quad (0,9), \quad (-2,6\frac{1}{3})$

24) $y = -\log_5(x + 15) - 6$ $\quad (10,-\frac{1}{5}), \quad (10,-8), \quad (11,-\frac{2}{5})$

25) $y = -3\log_2(2x + 6) + 7$ $\quad (5,-5), \quad (-5,-5), \quad (-2,-2)$

Properties of Logarithms

✍ **Expand each logarithm.**

1) $\log (11 \times 4) =$

2) $\log (13 \times 5) =$

3) $\log (4 \times 12) =$

4) $\log \left(\frac{2}{7}\right) =$

5) $\log \left(\frac{4}{9}\right) =$

6) $\log \left(\frac{5}{8}\right)^3 =$

7) $\log (6 \times 5^4) =$

8) $\log \left(\frac{14}{3}\right)^5 =$

9) $\log \left(\frac{3^4}{8}\right) =$

10) $\log (x \times y)^8 =$

11) $\log (x^6 \times y^{12} \times z^2) =$

12) $\log \left(\frac{u^8}{v^3}\right) =$

13) $\log \left(\frac{x}{y^7}\right) =$

✍ **Condense each expression to a single logarithm.**

14) $\log 8 - \log 13 =$

15) $\log 6 + \log 11 =$

16) $4 \log 2 - 7 \log 5 =$

17) $10 \log 4 - 3 \log 7 =$

18) $3 \log 9 - \log 17 =$

19) $11 \log 6 - 9 \log 4 =$

20) $\log 15 - 6 \log 7 =$

21) $6 \log 8 + 4 \log 10 =$

22) $12 \log 5 + 14 \log 9 =$

23) $17 \log_8 a + 6 \log_8 b =$

24) $2 \log_9 x - 3 \log_9 y =$

25) $\log_{11} u - 16 \log_{11} v =$

26) $8 \log_{15} u + 9 \log_{15} v =$

27) $32 \log_6 u - 25 \log_6 v =$

Natural Logarithms

✍ **Solve each equation for** x.

1) $e^x = 9$

2) $e^x = 36$

3) $e^x = 49$

4) $\ln x = 3$

5) $\ln(\ln x) = 7$

6) $e^x = 4$

7) $\ln(5x + 2) = 1$

8) $\ln(7x + 4) = 3$

9) $\ln(9x + 5) = 4$

10) $\ln x = \frac{1}{9}$

11) $\ln 11x = e^5$

12) $\ln x = \ln 6 + \ln 7$

13) $\ln x = 4\ln 3 + \ln 2$

✍ **Evaluate without using a calculator.**

14) $11\ln e =$

15) $\ln e^{10} =$

16) $4\ln e =$

17) $\ln e^{21} =$

18) $32\ln e =$

19) $4\ln e^5 =$

20) $e^{\ln 22} =$

21) $e^{3\ln 3} =$

22) $e^{3\ln 5} =$

23) $\ln \sqrt[11]{e} =$

✍ **Reduce the following expressions to simplest form.**

24) $e^{-4\ln 9 + 4\ln 3} =$

25) $e^{-3\ln\left(\frac{5}{4e}\right)} =$

26) $2\ln(e^4) =$

27) $\ln\left(\frac{1}{e}\right)^4 =$

28) $e^{\ln 9 + 3\ln 3} =$

29) $e^{\ln\left(\frac{13}{e}\right)} =$

30) $8\ln(1^{-3e}) =$

31) $2\ln\left(\frac{1}{e}\right)^{-3} =$

32) $6\ln\left(\frac{\sqrt[3]{e}}{3e}\right) =$

33) $e^{-4\ln e + 2\ln 5} =$

34) $e^{\ln\frac{4}{e}} =$

35) $11\ln(e^e) =$

ACT Subject Test – Mathematics

Exponential Equations and Logarithms

✎ **Solve each equation for the unknown variable.**

1) $3^{4n} = 243$

2) $5^{3r} = 625$

3) $6^{2n-1} = 216$

4) $16^{2r+3} = 4$

5) $169^{2x} = 13$

6) $7^{-3v-3} = 49$

7) $2^{4n} = 128$

8) $11^{n-1} = 1,331$

9) $\dfrac{9^{3a}}{3^{2a}} = 729$

10) $13^5 \times 13^{-4v} = 169$

11) $4^{3n} = \dfrac{1}{64}$

12) $(\dfrac{1}{11})^{2n} = 121$

13) $2{,}187^{3x} = 3$

14) $13^{5-7x} = 13^{-2x}$

15) $11^{-3x} = 11^{2x-7}$

16) $3^{5n} = 243$

17) $17^{5x+3} = 17^{6x}$

18) $15^{3n} = 225$

19) $4^{-3k} = 512$

20) $8^{-4r} = 8^{-5r+2}$

21) $8^{2x+3} = 8^{5x}$

22) $10^{3x-2} = 100{,}000$

23) $16 \times 64^{-v} = 128$

24) $\dfrac{128}{2^{-3m}} = 2^{4m+5}$

25) $14^{-5n} \times 14^{2n+3} = 14^{-2n}$

26) $(\dfrac{1}{9})^{4n+3} \times (\dfrac{1}{9})^{-3n-8} = (\dfrac{1}{9})^{-4n}$

✎ **Solve each problem. (Round to the nearest whole number)**

27) A substance decays 16% each day. After 8 days, there are 6 milligrams of the substance remaining. How many milligrams were there initially? _____

28) A culture of bacteria grows continuously. The culture doubles every 4 hours. If the initial number of bacteria is 20, how many bacteria will there be in 13 hours? _____

29) Bob plans to invest $11,200 at an annual rate of 3.5%. How much will Bob have in the account after three years if the balance is compounded quarterly? _____

30) Suppose you plan to invest $8,000 at an annual rate of 5%. How much will you have in the account after 6 years if the balance is compounded monthly? _____

Solving Logarithmic Equations

✍ **Find the value of the variables in each equation.**

1) $2\log(x) + 5 = 9$

2) $\log_4 4x + 3 = 5$

3) $-\log_8(8x) + 2 = 3$

4) $\log 2x - \log 4 = 1$

5) $\log 5x + \log 25 = 1$

6) $\log 4 - \log x = 3$

7) $\log 4x + \log 2 = \log 16$

8) $-6\log_3(5x - 1) = -36$

9) $\log 4x = \log(8x - 1)$

10) $\log(4k - 6) = \log(k - 3)$

11) $\log(5p + 2) = \log(p + 4)$

12) $-30 + \log_4(3n + 2) = -30$

13) $\log_4(4x - 4) = \log_4(x^2)$

14) $\log_8(k^2 + 15) = \log_8(-6k - 3)$

15) $\log(16 + 6b) = \log(10b^2 + 12b)$

16) $\log_6(2x + 5) - \log_6 x = \log_6 9$

17) $\log_5 5 + \log_5(x^2 + 1) = \log_5 25$

18) $\log_6(x + 3) + \log_6(x + 1) =$

$\log_6 8$

✍ **Find the value of x in each natural logarithm equation.**

19) $\ln 8 - \ln(4x + 8) = 4$

20) $\ln(x + 5) - \ln(x + 2) = \ln 10$

21) $\ln e^6 - \ln(x - 1) = 3$

22) $\ln(2x - 8) + \ln(x - 4) = \ln 8$

23) $\ln 5x - \ln(x + 4) = \ln 2$

24) $\ln(8x - 4) - \ln(x - 2) = \ln 25$

25) $\ln(3x + 2) - 4\ln 2 = 5$

26) $\ln(2x - 5) + \ln(x - 3) = \ln 6$

27) $\ln(x - 1) + \ln(4x - 7) = \ln(7)$

28) $2\ln 3x - \ln(x + 10) = \ln 2x$

29) $\ln x^4 + \ln x^8 = 4\ln(2x)$

30) $\ln x^{10} - \ln(x^2 + 10) = 10\ln 2x$

31) $8\ln(x - 2) = 4\ln(x^2 - 4x + 4)$

32) $\ln(x^4 + 10) = \ln(x^2 + 9)$

33) $2\ln x - 2\ln(x + 8) = \ln(x^2)$

34) $\ln(2x + 1) - \ln(4x + 1) = \ln 4$

35) $\ln 16 + 2\ln(x - 2) = \ln 4$

36) $\ln e^2 + \ln(5x - 6) = \ln(5) + 3$

Answers of Worksheets

Rewriting Logarithms

1) $3^3 = 27$

2) $2^7 = 128$

3) $6^4 = 1,296$

4) $5^4 = 625$

5) $11^2 = 121$

6) $12^3 = 1,728$

7) $9^3 = 729$

8) $3^6 = 729$

9) $10^4 = 10,000$

10) $7^3 = 343$

11) $4^5 = 1,024$

12) $12^2 = 144$

13) $13^3 = 2,197$

14) $25^{\frac{1}{2}} = 5$

15) $81^{\frac{1}{4}} = 3$

16) $3,125^{\frac{1}{5}} = 5$

17) $1,000^{\frac{1}{3}} = 10$

18) $5^{-3} = \frac{1}{125}$

19) $4^{-2} = \frac{1}{16}$

20) $a^b = \frac{7}{4}$

21) $\log_2 32 = 5$

22) $\log_4 64 = 3$

23) $\log_5 625 = 4$

24) $\log_{11} 1,331 = 3$

25) $\log_3 243 = 5$

26) $\log_6 1,296 = 4$

27) $\log_7 2,401 = 4$

28) $\log_9 729 = 3$

29) $\log_4 \frac{1}{1,024} = -5$

30) $\log_3 \frac{1}{6,561} = -8$

31) $\log_{11} \frac{1}{121} = -2$

32) $\log_{12} \frac{1}{1,728} = -3$

33) $\log_4 \frac{1}{1,024} = -5$

34) $\log_{10} \frac{1}{100,000} = -5$

Evaluating Logarithms

1) 6

2) 8

3) 5

4) 3

5) 2

6) 2

7) 4

8) -2

9) -4

10) -5

11) -2

12) -2

13) -2

14) $-\frac{1}{3}$

15) $\frac{1}{5}$

16) -2

17) -3

18) -4

19) $(2,5)$

20) $(\frac{1}{2}, -2)$

21) $(8, -1)$

22) $(6, 1)$

23) $(-2, 6\frac{1}{3})$

24) $(10, -8)$

25) $(5, -5)$

Properties of Logarithms

1) $\log 11 + \log 4$

2) $\log 13 + \log 5$

3) $\log 4 + \log 12$

4) $\log 2 - \log 7$

5) $\log 4 - \log 9$

6) $3 \log 5 - 3 \log 8$

7) $\log 6 + 4 \log 5$

8) $5 \log 14 - 5 \log 3$

9) $4 \log 3 - \log 8$

10) $8 \log x + 8 \log y$

11) $6\log x + 12\log y + 2\log z$

12) $8\log u - 3\log v$

13) $\log x - 7\log y$

14) $\log \frac{8}{13}$

15) $\log(6 \times 11)$

16) $\log \frac{2^4}{5^7}$

17) $\log \frac{4^{10}}{7^3}$

18) $\log \frac{9^3}{17}$

19) $\log \frac{6^{11}}{4^9}$

20) $\log \frac{15}{7^6}$

21) $\log (8^6 \times 10^4)$

22) $\log (5^{12} \times 9^{14})$

23) $\log_8 (a^{17}b^6)$

24) $\log_9 \frac{x^2}{y^3}$

25) $\log_{11} \frac{u}{v^{16}}$

26) $\log_{15}(u^8 \times v^9)$

27) $\log_6 \frac{u^{32}}{v^{25}}$

Natural Logarithms

1) $x = \ln 9$

2) $x = \ln 36, x = 2\ln (6)$

3) $x = \ln 49, x = 2\ln (7)$

4) $x = e^3$

5) $x = e^{e^7}$

6) $x = \ln 4$

7) $x = \frac{e-2}{5}$

8) $x = \frac{e^3-4}{7}$

9) $x = \frac{e^4-5}{9}$

10) $x = \sqrt[9]{e}$

11) $x = \frac{e\,e^5}{11}$

12) $x = 42$

13) $x = 162$

14) 11

15) 10

16) 4

17) 21

18) 32

19) 20

20) 22

21) 27

22) 125

23) $\frac{1}{11}$

24) $\frac{1}{81}$

25) $\frac{64e^3}{125}$

26) 8

27) -4

28) 243

29) $\frac{13}{e}$

30) 0

31) 6

32) $\ln \left(\frac{1}{3^6e^4}\right) = -10.6$

33) $25e^{-4} = \frac{25}{e^4}$

34) $\frac{4}{e}$

35) $11e$

Exponential Equations and Logarithms

1) $\frac{5}{4}$

2) $\frac{4}{3}$

3) 2

4) $-\frac{5}{4}$

5) $\frac{1}{4}$

6) $-\frac{5}{3}$

7) $\frac{7}{4}$

8) 4

ACT Subject Test – Mathematics

9) $\frac{3}{2}$

10) $\frac{3}{4}$

11) -1

12) -1

13) $\frac{1}{21}$

14) 1

15) $\frac{7}{5}$

16) 1

17) 3

18) $\frac{2}{3}$

19) $-\frac{3}{2}$

20) 2

21) 1

22) $\frac{7}{3}$

23) $-\frac{1}{2}$

24) 2

25) 3

26) 1

27) 24.2

28) 190.27

29) $\$12,432.4$

30) $\$10,792.14$

Solving Logarithmic Equations

1) $\{100\}$

2) $\{4\}$

3) $\{\frac{1}{64}\}$

4) $\{20\}$

5) $\{\frac{2}{25}\}$

6) $\{\frac{1}{250}\}$

7) $\{2\}$

8) $\{146\}$

9) $\{\frac{1}{4}\}$

10) No Solution

11) $\{\frac{1}{2}\}$

12) $\{-\frac{1}{3}\}$

13) $\{2\}$

14) No Solution

15) $\{1, -\frac{8}{5}\}$

16) $\{\frac{5}{7}\}$

17) $\{2, -2\}$

18) $\{1\}$

19) $x = \frac{8-8e^4}{4e^4}$

20) $\{-\frac{5}{3}\}$

21) $e^3 + 1$

22) $\{6\}$

23) $\{\frac{8}{3}\}$

24) $\{\frac{46}{17}\}$

25) $x = \frac{16e^5-2}{3}$

26) $x = \frac{9}{2}$

27) $x = \frac{11}{4}$

28) $x = \frac{20}{7}$

29) $e^{\frac{\ln(2)}{2}}$

30) No Solution

31) $x > 2$

32) No Solution

33) No Solution

34) $x = -\frac{3}{14}$

35) $x = \frac{5}{2}$

36) $x = \frac{5e+6}{5}$

Chapter 13 :
Geometry and Solid Figures

Topics that you'll practice in this chapter:

- ✓ Angles
- ✓ Pythagorean Relationship
- ✓ Triangles
- ✓ Polygons
- ✓ Trapezoids
- ✓ Circles
- ✓ Cubes
- ✓ Rectangular Prism
- ✓ Cylinder
- ✓ Pyramids and Cone

Geometry is the archetype of the beauty of the world.

Johannes Kepler

Angles

✑ **What is the value of x in the following figures?**

1)
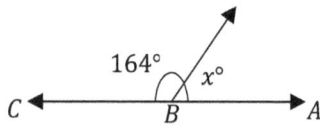

2)
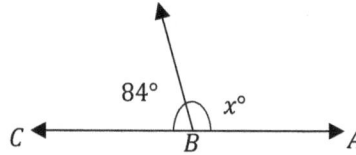

3)
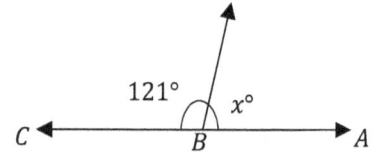

4)
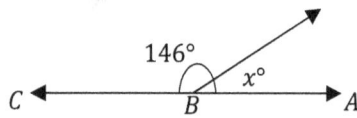

5)
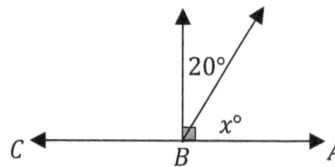

6)
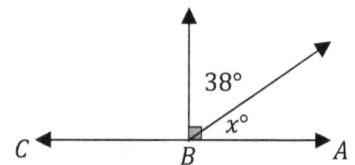

✑ **Calculate.**

7) Two supplement angles have equal measures. What is the measure of each angle? _____

8) The measure of an angle is seven fifth the measure of its supplement. What is the measure of the angle? _____

9) Two angles are complementary and the measure of one angle is 24 less than the other. What is the measure of the smaller angle? _____

10) Two angles are complementary. The measure of one angle is one fifth the measure of the other. What is the measure of the bigger angle? _____

11) Two supplementary angles are given. The measure of one angle is 40° less than the measure of the other. What does the smaller angle measure? _____

Pythagorean Relationship

✍ Do the following lengths form a right triangle?

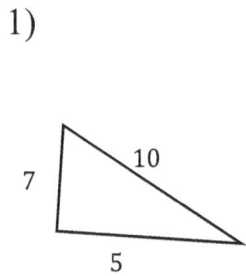

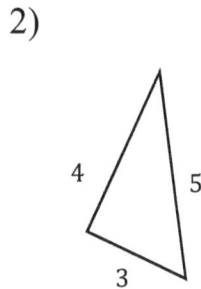

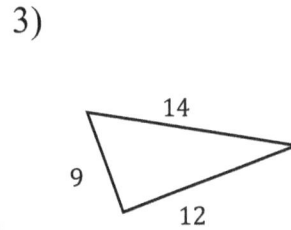

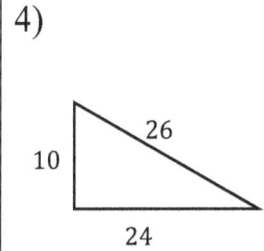

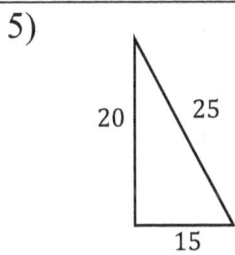

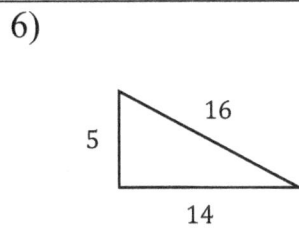

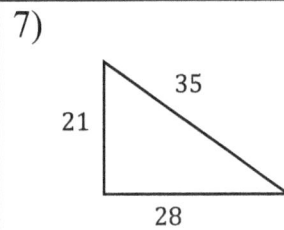

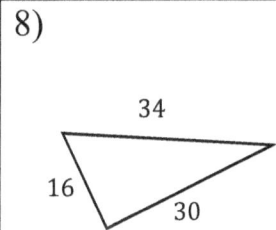

1)

7   10   5

2)

4   5   3

3)

14   9   12

4)

10   26   24

5)

20   25   15

6)

5   16   14

7)

21   35   28

8)

34   16   30

✍ Find the missing side?

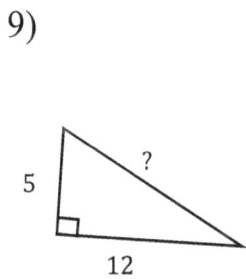

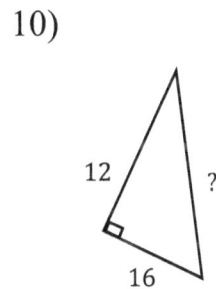

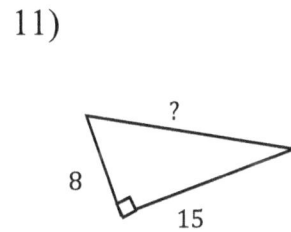

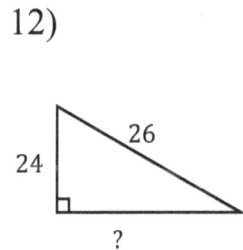

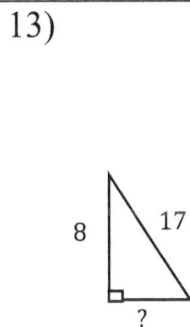

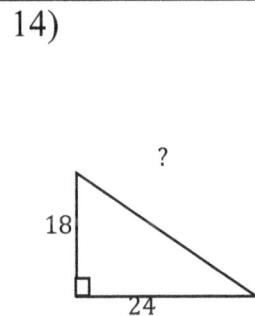

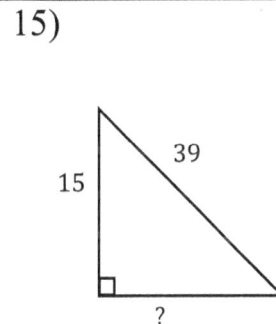

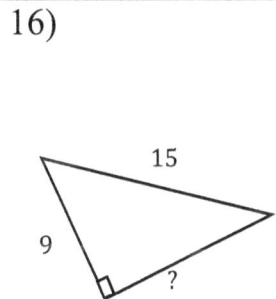

9)

5   ?   12

10)

12   ?   16

11)

?   8   15

12)

24   26   ?

13)

8   17   ?

14)

18   ?   24

15)

15   39   ?

16)

15   9   ?

Triangles

✏️ **Find the measure of the unknown angle in each triangle.**

1)
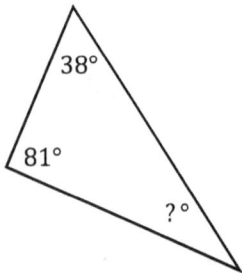
38°
81°
? °

2)
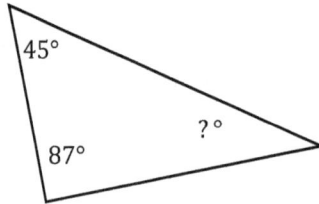
45°
87°
? °

3)
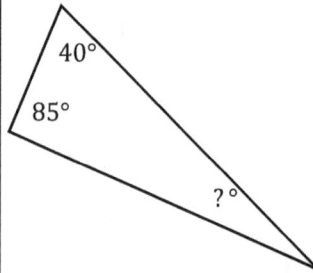
40°
85°
? °

4)
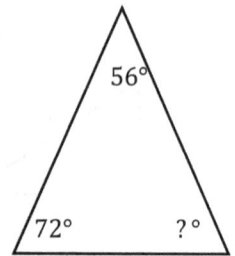
56°
72°
? °

5)
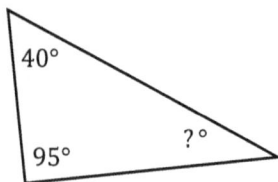
40°
95°
? °

6)
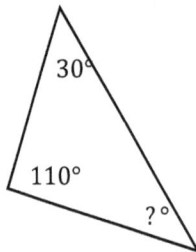
30°
110°
? °

7)
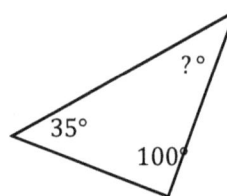
? °
35°
100°

8)
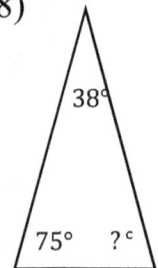
38°
75°
? °

✏️ **Find area of each triangle.**

9)
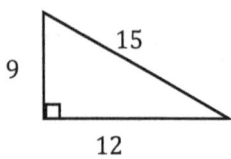
15
9
12

10)
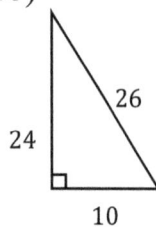
26
24
10

11)
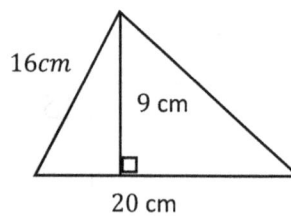
16cm
9 cm
20 cm

12)
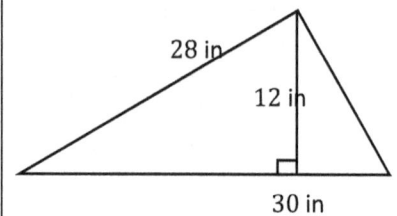
28 in
12 in
30 in

Polygons

✍ **Find the perimeter of each shape.**

1)

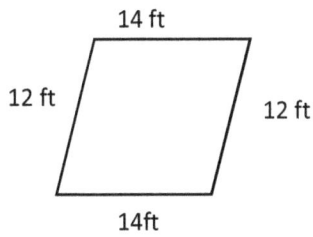

14 ft

12 ft   12 ft

14ft

2)

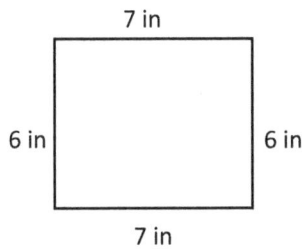

7 in

6 in   6 in

7 in

3)

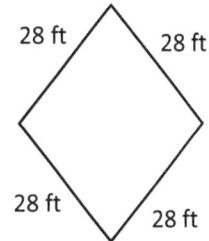

28 ft   28 ft

28 ft

28 ft

4) Square

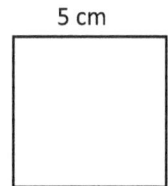

5 cm

5) Regular hexagon

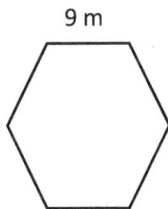

9 m

6)

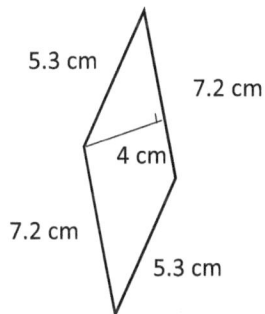

5.3 cm

7.2 cm

4 cm

7.2 cm

5.3 cm

7) Parallelogram

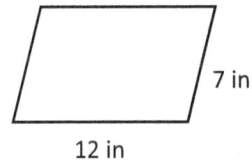

7 in

12 in

8) Square

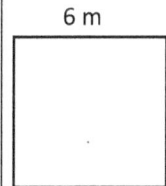

6 m

✍ **Find the area of each shape.**

9) Parallelogram

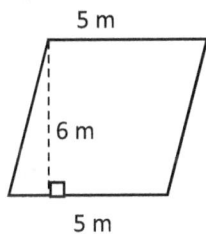

5 m

6 m

5 m

10) Rectangle

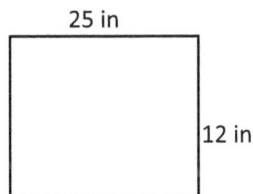

25 in

12 in

11) Rectangle

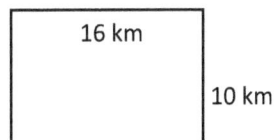

16 km

10 km

12) Square

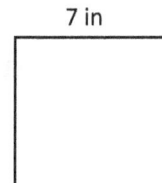

7 in

Trapezoids

✎ **Find the area of each trapezoid.**

1)

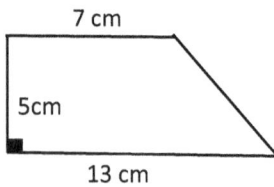

7 cm
5cm
13 cm

2)

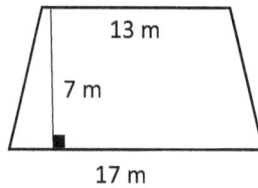

13 m
7 m
17 m

3)

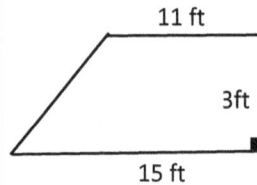

11 ft
3ft
15 ft

4)

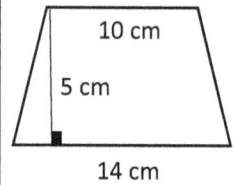

10 cm
5 cm
14 cm

5)

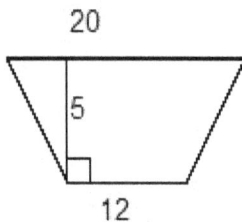

20
5
12

6)

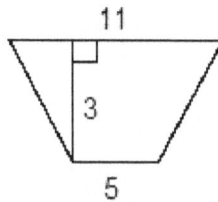

11
3
5

7)

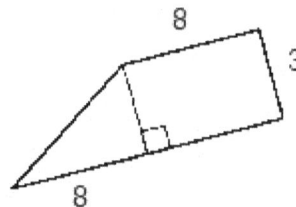

8
3
8

8)

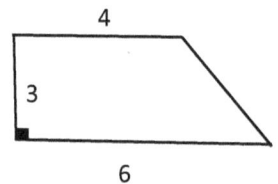

4
3
6

✎ **Calculate.**

1) A trapezoid has an area of 45 cm² and its height is 5 cm and one base is 5 cm. What is the other base length? _____

2) If a trapezoid has an area of 99 ft² and the lengths of the bases are 8 ft and 10 ft, find the height? _____

3) If a trapezoid has an area of 126 m² and its height is 14 m and one base is 6 m, find the other base length? _____

4) The area of a trapezoid is 440 ft² and its height is 22 ft. If one base of the trapezoid is 15 ft, what is the other base length?

_____

Circles

✎ **Find the area of each circle.** ($\pi = 3.14$)

1)	2)	3)	4)	5)	6)
2.5 in	5cm	9 ft	2 m	11 cm	10 miles

7)	8)	9)	10)	11)	12)
13 in	3 ft	4 m	12 cm	7 miles	8 ft

✎ **Complete the table below.** ($\pi = 3.14$)

Circle No.	Radius	Diameter	Circumference	Area
1	1 in	2 in	6.28 in	3.14 in^2
2		10 m		
3				28.26 ft^2
4			47.1 mi	
5		11 km		
6	7 cm			
7		12 ft		
8				314 m^2
9			56.52 in	
10	4.5 ft			

Cubes

✎ **Find the volume of each cube.**

1)	2)	3)	4)	5)	6)
	2 cm	6 ft	11 m	13 in	7 miles

7)	8)	9)	10)	11)	12)
1.2 km	9 cm	2.1 ft	12 mm	0.2 in	0.1 km

✎ **Find the surface area of each cube.**

13)	14)	15)	16)	17)	18)
	7 m	5 ft	4.5 mm	1.1 km	11 cm

Rectangular Prism

✎ **Find the volume of each Rectangular Prism.**

1)

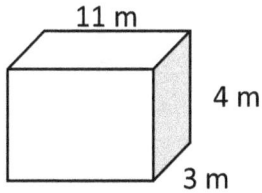

11 m
4 m
3 m

2)

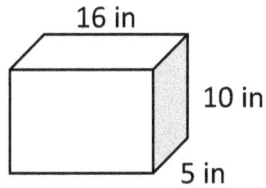

16 in
10 in
5 in

3)

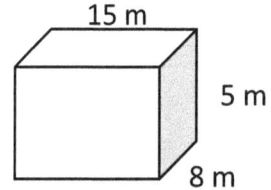

15 m
5 m
8 m

4)

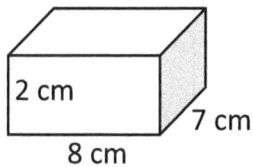

2 cm
7 cm
8 cm

5)

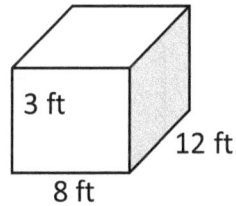

3 ft
12 ft
8 ft

6)

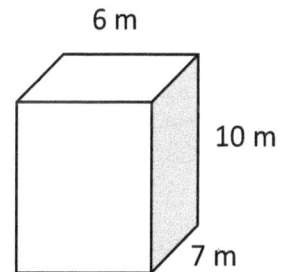

6 m
10 m
7 m

✎ **Find the surface area of each Rectangular Prism.**

7)

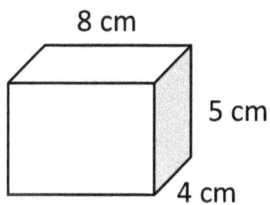

8 cm
5 cm
4 cm

8)

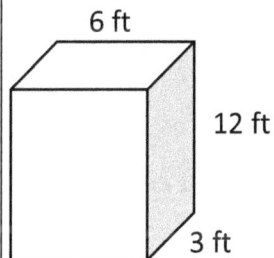

6 ft
12 ft
3 ft

9)

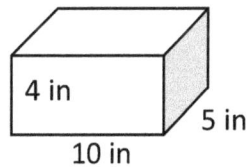

4 in
10 in
5 in

10)

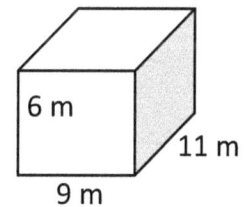

6 m
11 m
9 m

Cylinder

✎ **Find the volume of each Cylinder. Round your answer to the nearest tenth.** ($\pi = 3.14$)

1)

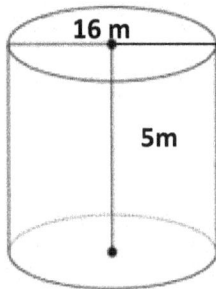

16 m

5m

2)

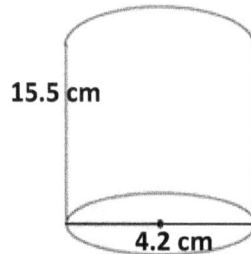

15.5 cm

4.2 cm

3)

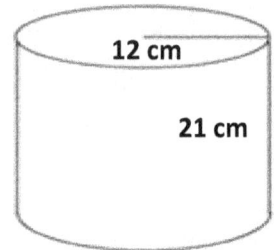

12 cm

21 cm

4)

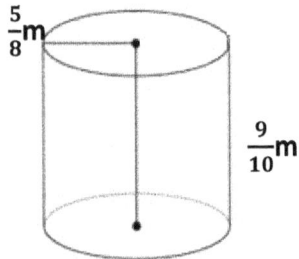

$\frac{5}{8}$m

$\frac{9}{10}$m

5)

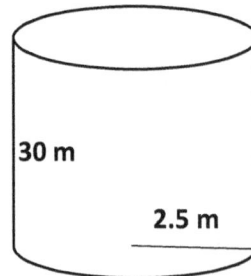

30 m

2.5 m

6)

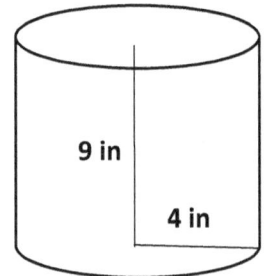

9 in

4 in

✎ **Find the surface area of each Cylinder.** ($\pi = 3.14$)

7)

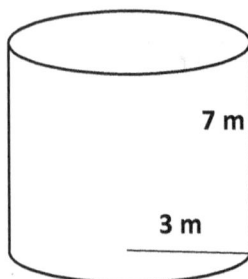

7 m

3 m

8)

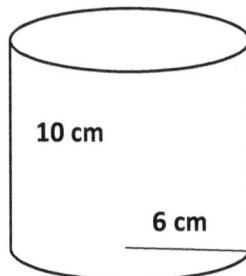

10 cm

6 cm

9)

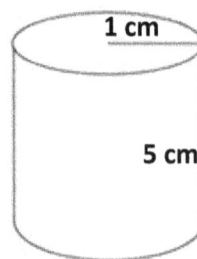

1 cm

5 cm

10)

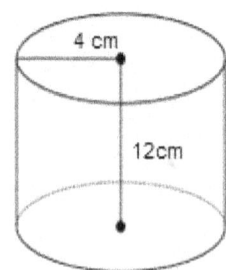

4 cm

12cm

Pyramids and Cone

🖎 **Find the volume of each Pyramid and Cone.** ($\pi = 3.14$)

1)

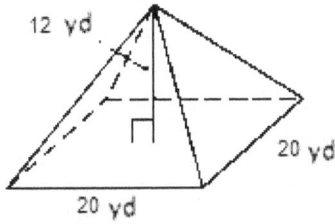

2)

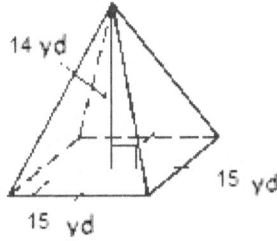

3)

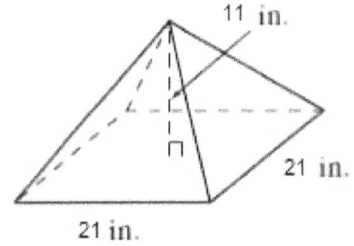

4)

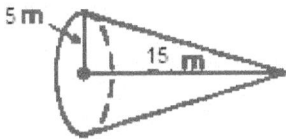

5)

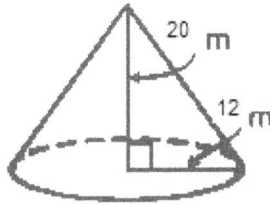

6)

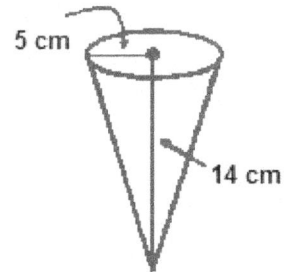

🖎 **Find the surface area of each Pyramid and Cone.** ($\pi = 3.14$)

7)   8)   9)   10)

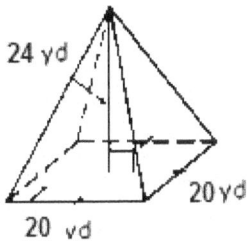
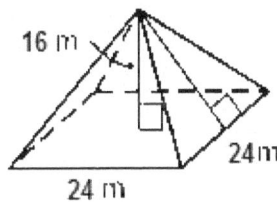
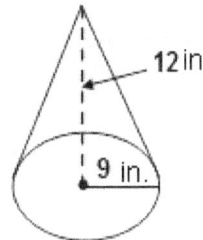
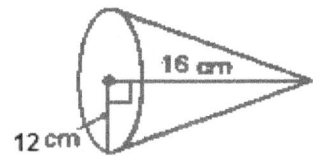

Answers of Worksheets

Angles

1) 16°	4) 34°	7) 90°	10) 75°
2) 96°	5) 70°	8) 75°	11) 70°
3) 59°	6) 52°	9) 33°	

Pythagorean Relationship

1) No	5) Yes	9) 13	13) 15
2) Yes	6) No	10) 20	14) 30
3) No	7) Yes	11) 17	15) 36
4) Yes	8) Yes	12) 10	16) 12

Triangles

1) 60°	5) 45°	9) 54 *square unites*
2) 48°	6) 40°	10) 120 *square unites*
3) 55°	7) 45°	11) 90 *square unites*
4) 52°	8) 67°	12) 180 *square unites*

Polygons

1) 52 ft	5) 54 m	9) 30 m^2
2) 26 in	6) 25 cm	10) 300 in^2
3) 112 ft	7) 38 in	11) 160 km^2
4) 20 cm	8) 24 m	12) 49 in^2

Trapezoids

1) 50 cm^2	4) 60 cm^2	7) 36
2) 105 m^2	5) 80	8) 15
3) 39 ft^2	6) 24	

Calculate

1) 13 cm	2) 11 ft	3) 12 m	4) 25 ft

Circles

1) 19.63 in^2	5) 379.94 cm^2	9) 12.56 m^2
2) 78.5 cm^2	6) 314 $miles^2$	10) 113.04 cm^2
3) 254.34 ft^2	7) 132.67 in^2	11) 38.47 $miles^2$
4) 12.56 m^2	8) 7.07 ft^2	12) 50.24 ft^2

Circle No.	Radius	Diameter	Circumference	Area
1	1 in	2 in	6.28 in	3.14 in^2
2	5 m	10 m	31.4 m	78.5 m^2
3	3 ft	6 ft	18.84 ft	28.26 ft^2
4	7.5 miles	15 mi	47.1 mi	176.63 mi^2
5	5.5 km	11 km	34.54 km	94.99 km^2
6	7 cm	14 cm	43.96 cm	153.86 cm^2
7	6 ft	12 ft	37.68 feet	113.04 ft^2
8	10 m	20 m	62.8 m	314 m^2
9	9 in	18 in	56.52 in	254.34 in^2
10	4.5 ft	9 ft	28.26 ft	63.585 ft^2

Cubes

1) 12
2) 8 cm^3
3) 216 ft^3
4) 1,331 m^3
5) 2,197 in^3
6) 343 $miles^3$
7) 1.728 km^3
8) 729 cm^3
9) 9.261 ft^3
10) 1,728 mm^3
11) 0.008 in^3
12) 0.001 km^3
13) 27
14) 294 m^2
15) 150 ft^2
16) 121.5 mm^2
17) 7.26 km^2
18) 726 cm^2

Rectangular Prism

1) 132 m^3
2) 800 in^3
3) 600 m^3
4) 112 cm^3
5) 288 ft^3
6) 420 m^3
7) 184 cm^2
8) 252 ft^2
9) 220 in^2
10) 438 m^2

Cylinder

1) 1,004.8 m^3
2) 214.6 cm^3
3) 9,495.4 cm^3
4) 1.1 m^3
5) 588.8 m^3
6) 452.2 in^3
7) 188.4 m^2
8) 602.9 cm^2
9) 37.7 cm^2
10) 401.9 m^2

Pyramids and Cone

1) 1,600 yd^3
2) 1,050 yd^3
3) 1,617 in^3
4) 392.5 m^3
5) 3,014.4 m^3
6) 366.33 cm^3
7) 1,440 yd^2
8) 1,536 m^2
9) 678.24 in^2
10) 1,205.76 cm^2

Chapter 14 :

Trigonometric Functions

Topics that you'll practice in this chapter:

- ✓ Trig ratios of General Angles
- ✓ Sketch Each Angle in Standard Position
- ✓ Finding Co–Terminal Angles and Reference Angles
- ✓ Angles in Radians
- ✓ Angles in Degrees
- ✓ Evaluating Each Trigonometric Expression
- ✓ Missing Sides and Angles of a Right Triangle
- ✓ Arc Length and Sector Area

Mathematics is like checkers in being suitable for the young, not too difficult, amusing, and without peril to the state. — *Plato*

Trig ratios of General Angles

✎ **Evaluate.**

1) $\sin 135° = $ _____

2) $\sin 300° = $ _____

3) $\cos - 225° = $ _____

4) $\cos 270° = $ _____

5) $\sin 450° = $ _____

6) $\sin -330° = $ _____

7) $\tan 60° = $ _____

8) $\cot 180° = $ _____

9) $\tan 240° = $ _____

10) $\cot 90° = $ _____

11) $\sec 180° = $ _____

12) $\csc 90° = $ _____

13) $\cot -270° = $ _____

14) $\sec 360° = $ _____

15) $\cos - 45° = $ _____

16) $\sec 120° = $ _____

17) $\csc 360° = $ _____

18) $\cot -45° = $ _____

✎ **Find the exact value of each trigonometric function. Some may be undefined.**

19) $\sec 2\pi = $ _____

20) $\tan -\dfrac{5\pi}{2} = $ _____

21) $\cos \dfrac{11\pi}{2} = $ _____

22) $\cot \dfrac{9\pi}{4} = $ _____

23) $\sec - 6\pi = $ _____

24) $\sec \dfrac{\pi}{4} = $ _____

25) $\csc \dfrac{8\pi}{3} = $ _____

26) $\cot \dfrac{10\pi}{3} = $ _____

27) $\csc -\dfrac{\pi}{2} = $ _____

28) $\cot \dfrac{2\pi}{3} = $ _____

Sketch Each Angle in Standard Position

✎ **Draw each angle with the given measure in standard position.**

1) −570°

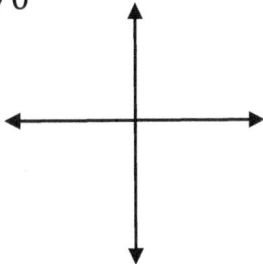

4) −690°

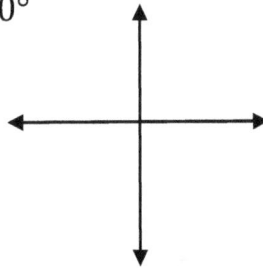

2) 750°

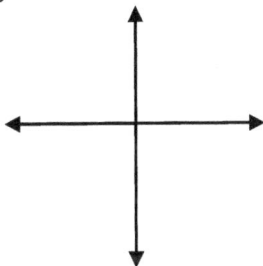

5) $\frac{13\pi}{6}$

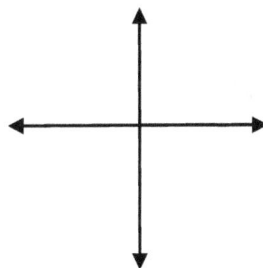

3) 1,110°

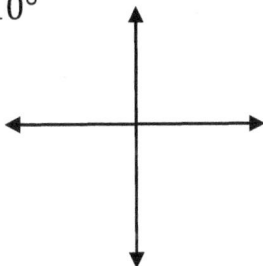

6) $-\frac{11\pi}{6}$

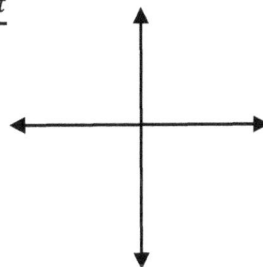

Finding Co-terminal Angles and Reference Angles

✎ **Find a conterminal angle between 0° and 360° for each angle provided.**

1) $-315° =$

3) $-225° =$

2) $-210° =$

4) $-540° =$

✎ **Find a conterminal angle between 0 and 2π for each given angle.**

5) $\dfrac{18\pi}{5} =$

7) $-\dfrac{13\pi}{4} =$

6) $-\dfrac{19\pi}{6} =$

8) $\dfrac{14\pi}{3} =$

✎ **Find the reference angle of each angle.**

9)

10)

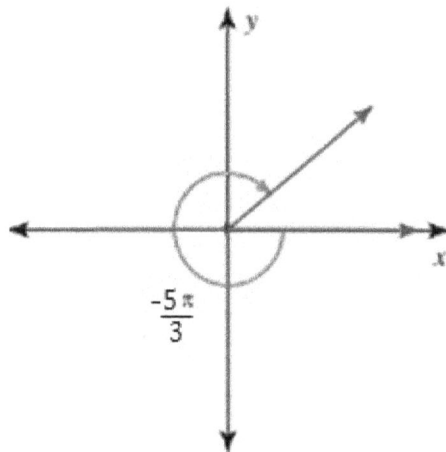

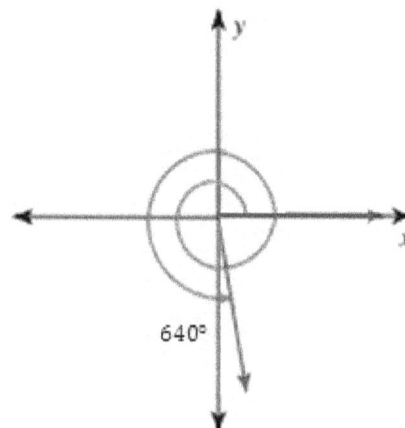

Angles and Angle Measure

✍ **Convert each degree measure into radians.**

1) $216° = $ ____

2) $660° = $ ____

3) $420° = $ ____

4) $220° = $ ____

5) $210° = $ ____

6) $270° = $ ____

7) $-300° = $ ____

8) $810° = $ ____

9) $330° = $ ____

10) $140° = $ ____

11) $480° = $ ____

12) $405° = $ ____

13) $-450° = $ ____

14) $-126° = $ ____

15) $-675° = $ ____

16) $150° = $ ____

17) $-468° = $ ____

18) $340° = $ ____

19) $-440° = $ ____

20) $342° = $ ____

21) $230° = $ ____

✍ **Convert each radian measure into degrees.**

22) $\frac{\pi}{10} = $

23) $\frac{5\pi}{12} = $

24) $\frac{7\pi}{3} = $

25) $\frac{3\pi}{20} = $

26) $-\frac{6\pi}{5} = $

27) $\frac{11\pi}{18} = $

28) $-\frac{14\pi}{5} = $

29) $\frac{5\pi}{18} = $

30) $\frac{7\pi}{36} = $

31) $\frac{17\pi}{18} = $

32) $-\frac{13\pi}{30} = $

33) $\frac{7\pi}{9} = $

34) $-\frac{19\pi}{18} = $

35) $\frac{7\pi}{60} = $

36) $-\frac{3\pi}{10} = $

37) $\frac{11\pi}{30} = $

38) $-\frac{2\pi}{9} = $

39) $-\frac{7\pi}{10} = $

Evaluating Trigonometric Functions

✍ **Find the exact value of each trigonometric function.**

1) $\cos 780° = $ _____

2) $\tan \dfrac{5\pi}{3} = $ _____

3) $\tan -\dfrac{\pi}{6} = $ _____

4) $\cot -\dfrac{9\pi}{4} = $ _____

5) $\cos -\dfrac{7\pi}{6} = $ _____

6) $\cos 135° = $ _____

7) $\sin 240° = $ _____

8) $\tan 330° = $ _____

9) $\cot 420° = $ _____

10) $\tan - 495° = $ _____

11) $\cot 315° = $ _____

12) $\sin - 240° = $ _____

13) $\cot 225° = $ _____

✍ **Use the given point on the terminal side of angle θ to find the value of the trigonometric function indicated.**

14) $\sin \theta,\ (-6, 8)$

15) $\cos \theta,\ (-6, 8)$

16) $\sec \theta,\ (3,\ 5)$

17) $\cos \theta,\ (10, 24)$

18) $\sin \theta,\ (6, -6)$

19) $\tan \theta,\ (-2, -\sqrt{12})$

Missing Sides and Angles of a Right Triangle

✎ Find the value of each trigonometric ratio as fractions in their simplest form.

1) $cot\ x$

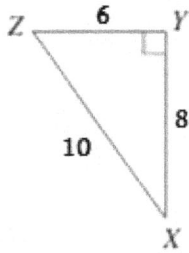

2) $cos\ A$

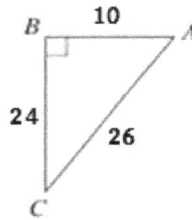

✎ Find the missing sides. Round answers to the nearest tenth.

3)

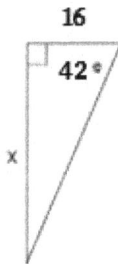

4)

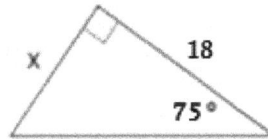

5)

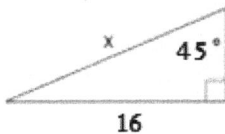

6)

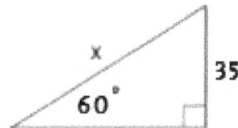

Arc Length and Sector Area

✎ **Find the length of each arc. Round your answers to the nearest tenth.**

$(\pi = 3.14)$

1) $r = 28$ cm, $\theta = 30°$

3) $r = 22$ ft, $\theta = 50°$

2) $r = 14$ ft, $\theta = 95°$

4) $r = 16\,m$, $\theta = 85°$

✎ **Find area of each sector. Do *not* round. Round your answers to the nearest tenth.** $(\pi = 3.14)$

5)

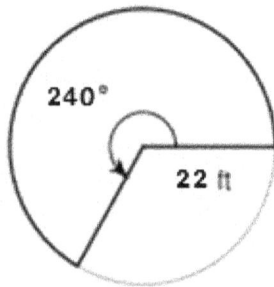

240°

22 ft

7)

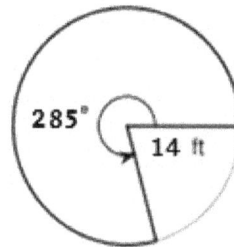

285°

14 ft

6)

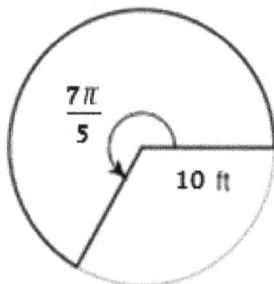

$\dfrac{7\pi}{5}$

10 ft

8)

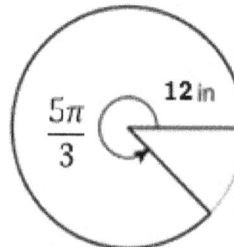

$\dfrac{5\pi}{3}$

12 in

Answers of Worksheets

Trig Ratios of General Angles

1) $\frac{\sqrt{2}}{2}$

2) $-\frac{\sqrt{3}}{2}$

3) $-\frac{\sqrt{2}}{2}$

4) 0

5) 1

6) $\frac{1}{2}$

7) $\sqrt{3}$

8) Undefined

9) $\sqrt{3}$

10) 0

11) -1

12) 1

13) 0

14) 1

15) $\frac{\sqrt{2}}{2}$

16) -2

17) Undefined

18) -1

19) 1

20) Undefined

21) 0

22) 1

23) 1

24) $\sqrt{2}$

25) $\frac{2\sqrt{3}}{3}$

26) $\frac{\sqrt{3}}{3}$

27) -1

28) $-\frac{\sqrt{3}}{3}$

Sketch Each Angle in Standard Position

1) $-570°$

2) $750°$

3) $1,110°$

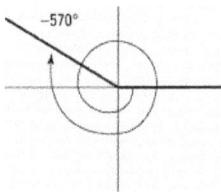
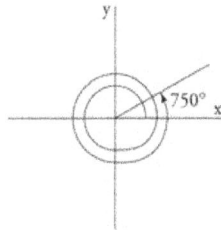
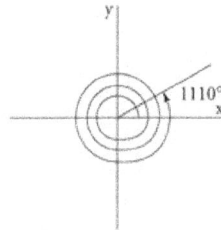

4) $-690°$

5) $\frac{13\pi}{6} = 390°$

6) $-\frac{11\pi}{6} = -330°$

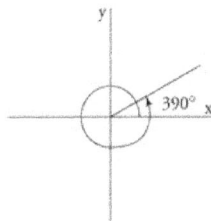
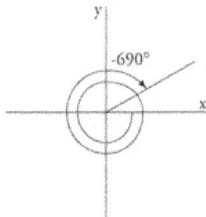
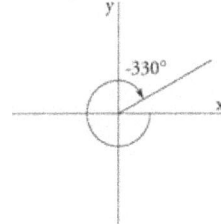

Finding Co–Terminal Angles and Reference Angles

1) $45°$

2) $150°$

3) $135°$

4) $180°$

5) $\frac{4\pi}{5}$

6) $\frac{5\pi}{6}$

7) $\frac{3\pi}{4}$

8) $\frac{2\pi}{3}$

9) $\frac{\pi}{3}$

10) $80°$

Angles and Angle Measure

1) $\frac{6\pi}{5}$

2) $\frac{11\pi}{3}$

3) $\frac{7\pi}{3}$

4) $\frac{11\pi}{9}$

5) $\frac{7\pi}{6}$

6) $\frac{3\pi}{2}$

7) $-\frac{5\pi}{3}$

8) $\frac{9\pi}{2}$

9) $\frac{11\pi}{6}$

10) $\frac{7\pi}{9}$

11) $\frac{8\pi}{3}$

12) $\frac{9\pi}{4}$

13) $-\frac{5}{2}\pi$

14) $-\frac{7\pi}{10}$

15) $-\frac{15\pi}{4}$

16) $\frac{5\pi}{6}$

17) $-\frac{13\pi}{5}$

18) $\frac{17\pi}{9}$

19) $-\frac{22\pi}{9}$

20) $\frac{19\pi}{10}$

21) $\frac{23\pi}{18}$

22) $18°$

23) $75°$

24) $420°$

25) $27°$

26) $-216°$

27) $110°$

28) $-504°$

29) $50°$

30) $35°$

31) $170°$

32) $-78°$

33) $140°$

34) $-190°$

35) $21°$

36) $-54°$

37) $66°$

38) $-40°$

39) $-126°$

Evaluating Each Trigonometric Functions

1) $\frac{1}{2}$

2) $-\sqrt{3}$

3) $-\frac{\sqrt{3}}{3}$

4) -1

5) $-\frac{\sqrt{3}}{2}$

6) $-\frac{\sqrt{2}}{2}$

7) $-\frac{\sqrt{3}}{2}$

8) $-\frac{\sqrt{3}}{3}$

9) $\frac{\sqrt{3}}{3}$

10) 1

11) -1

12) $\frac{\sqrt{3}}{2}$

13) 1

14) 0.8

15) -0.6

16) $\frac{\sqrt{34}}{5}$

17) $\frac{5}{13}$

18) $-\frac{\sqrt{2}}{2}$

19) $\sqrt{3}$

Missing Sides and Angles of a Right Triangle

1) $\frac{4}{3}$

2) $\frac{5}{13}$

3) 14.4

4) 67.2

5) 22.6

6) 40.4

Arc Length and Sector Area

1) $14.7\ cm$

2) $23.2\ ft$

3) $19.2\ ft$

4) $23.7m$

5) $1,013.7\ ft^2$

6) $220\ in^2$

7) $487.5\ ft^2$

8) $377\ in^2$

Chapter 15 :

Statistics and Probability

Topics that you'll practice in this chapter:

✓ Mean and Median

✓ Mode and Range

✓ Histograms

✓ Stem–and–Leaf Plot

✓ Pie Graph

✓ Probability Problems

✓ Factorials

✓ Combinations and Permutation

"The book of nature is written in the language of Mathematic."

- Galileo.

Mean and Median

✎ **Find Mean and Median of the Given Data.**

1) 8, 7, 14, 4, 8

2) 14, 8, 25, 19, 16, 33, 11

3) 23, 18, 15, 12, 17

4) 34, 14, 10, 15, 6, 11

5) 10, 19, 6, 8, 32, 20, 17

6) 17, 26, 39, 69, 20, 6

7) 40, 38, 18, 11, 9, 2, 7, 32, 41

8) 24, 21, 31, 12, 33, 32, 22

9) 16, 14, 20, 41, 15, 20, 38, 4

10) 20, 20, 30, 18, 6, 28, 12, 46

11) 12, 7, 10, 11, 16, 22

12) 10, 29, 27, 12, 2, 15, 10, 3

✎ **Calculate.**

13) In a javelin throw competition, five athletics score 56, 34, 62, 23 and 19 meters. What are their Mean and Median? _____

14) Eva went to shop and bought 8 apples, 14 peaches, 6 bananas, 4 pineapples and 12 melons. What are the Mean and Median of her purchase? _____

15) Bob has 17 black pen, 19 red pen, 14 green pens, 20 blue pens and 5 boxes of yellow pens. If the Mean and Median are 19 respectively, what is the number of yellow pens in each box? _____

Mode and Range

✎ **Find Mode and Rage of the Given Data.**

1) 4, 3, 7, 3, 3, 4

    Mode: _____   Range: _____

2) 18, 18, 24, 26, 18, 8, 14, 22

    Mode: _____   Range: _____

3) 8, 8, 8, 16, 19, 22, 20, 9, 13

    Mode: _____   Range: _____

4) 24, 24, 14, 28, 20, 18, 20, 24

    Mode: _____   Range: _____

5) 6, 21, 27, 24, 27, 27

    Mode: _____   Range: _____

6) 21, 8, 8, 7, 8, 12, 10, 22, 18, 13

    Mode: _____   Range: _____

7) 7, 4, 4, 6, 13, 13, 13, 0, 2, 2

    Mode: _____   Range: _____

8) 5, 8, 5, 14, 12, 14, 3, 5, 18

    Mode: _____   Range: _____

9) 7, 7, 7, 12, 7, 3, 8, 16, 3, 17

    Mode: _____   Range: _____

10) 15, 15, 19, 16, 4, 16, 10, 15

    Mode: _____   Range: _____

11) 6, 6, 5, 6, 42, 13, 19, 2

    Mode: _____   Range: _____

12) 8, 8, 9, 8, 9, 4, 34, 22

    Mode: _____   Range: _____

✎ **Calculate.**

13) A stationery sold 12 pencils, 56 red pens, 24 blue pens, 20 notebooks, 12 erasers, 21 rulers and 11 color pencils. What are the Mode and Range for the stationery sells?

                Mode: _____   Range: _____

14) In an English test, eight students score 10, 15, 15, 18 18, 16, 15 and 15. What are their Mode and Range? _____

15) What is the range of the first 6 even numbers greater than 8?

    _____

Times Series

🖎 **Use the following Graph to complete the table.**

Day	Distance (km)
1	
2	

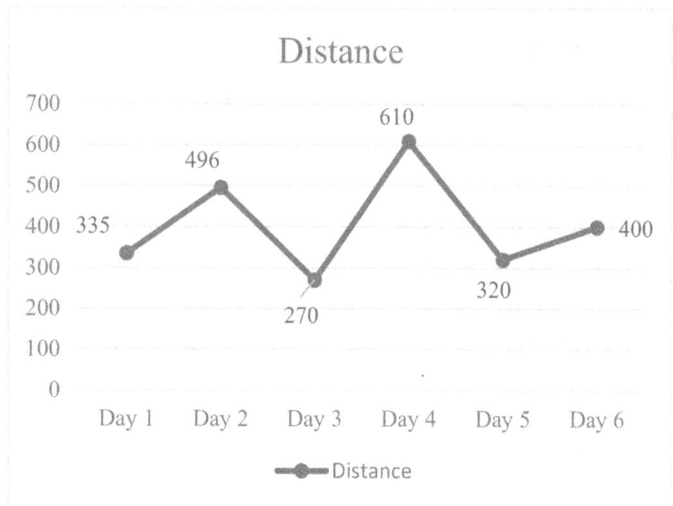

Distance

```
700
600                      610
500        496
400  335                              400
300
200                 270        320
100
  0
   Day 1  Day 2  Day 3  Day 4  Day 5  Day 6
              Distance
```

The following table shows the number of births in the US from 2007 to 2012 (in millions).

Year	Number of births (in millions)
2007	4.15
2008	3.70
2009	3.45
2010	3.20
2011	1.75
2012	2.98

Draw a Time Series for the table.

Stem–and–Leaf Plot

✎Make stem ad leaf plots for the given data.

1) 24, 26, 29, 20, 53, 27, 51, 55, 36, 21, 37, 30

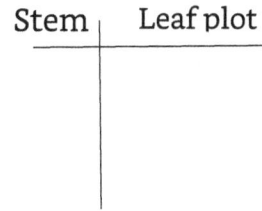

Stem | Leaf plot

2) 11, 59, 66, 14, 18, 19, 59, 65, 69, 61, 68, 65

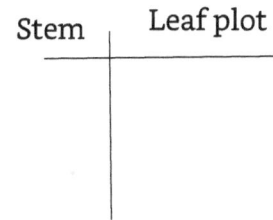

Stem | Leaf plot

3) 121, 55, 66, 54, 112, 128, 63, 125, 59, 123, 68, 119

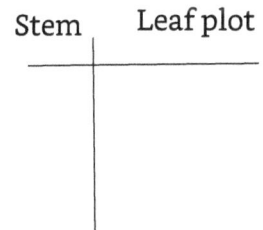

Stem | Leaf plot

4) 51, 32, 100, 56, 84, 36, 107, 56, 85, 39, 56, 106, 89

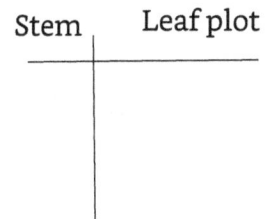

Stem | Leaf plot

5) 33, 89, 19, 87, 81, 16, 11, 30, 86, 35, 17, 35, 13

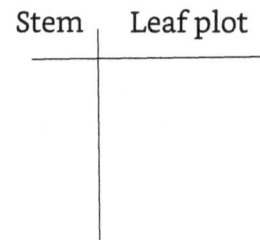

Stem | Leaf plot

6) 60, 92, 22, 25, 67, 93, 95, 62, 21, 64, 98, 29

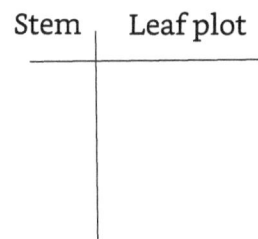

Stem | Leaf plot

Pie Graph

The circle graph below shows all Robert's expenses for last month. Robert spent $140 on his hobbies last month.

Answer following questions based on the Pie graph.

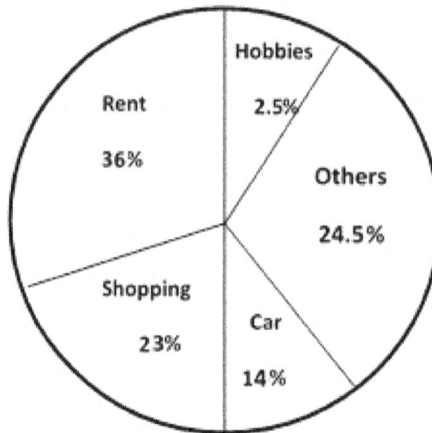

1) How much was Robert's total expenses last month? _____

2) How much did Robert spend on his car last month? _____

3) How much did Robert spend for shopping last month? _____

4) How much did Robert spend on his rent last month? _____

5) What fraction is Robert's expenses for his rent and car out of his total

   expenses last month? _____

Probability Problems

✎ **Calculate.**

1) A number is chosen at random from 1 to 10. Find the probability of selecting number 6 or smaller numbers. _____

2) Bag A contains 18 red marbles and 6 green marbles. Bag B contains 16 black marbles and 8 orange marbles. What is the probability of selecting a green marble at random from bag A? What is the probability of selecting a black marble at random from Bag B? _____

3) A number is chosen at random from 1 to 20. What is the probability of selecting multiples of 4? _____

4) A card is chosen from a well-shuffled deck of 52 cards. What is the probability that the card will be a queen? _____

5) A number is chosen at random from 1 to 15. What is the probability of selecting a multiple of 3 or 5? _____

A spinner numbered 1–8, is spun once. What is the probability of spinning …?

6) an Odd number? _____   7) a multiple of 2? _____

8) a multiple of 5? _____   9) number 10? _____

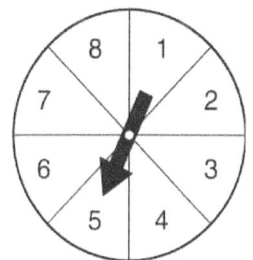

Factorials

✎ **Determine the value for each expression.**

1) $4! + 0! =$

2) $2! + 5! =$

3) $(2!)^2 =$

4) $5! - 3! =$

5) $6! - 3! + 10 =$

6) $3! \times 4 - 15 =$

7) $(2! + 3!)^2 =$

8) $(4! - 3!)^2 =$

9) $(3!\,0!)^2 - 10 =$

10) $\dfrac{10!}{8!} =$

11) $\dfrac{6!}{4!} =$

12) $\dfrac{6!}{5!} =$

13) $\dfrac{15!}{13!} =$

14) $\dfrac{n!}{(n-3)!} =$

15) $\dfrac{(n+2)!}{n!} =$

16) $\dfrac{(2+2!)^3}{2!} =$

17) $\dfrac{5(n+2)!}{(n+1)!} =$

18) $\dfrac{22!}{20!4!} =$

19) $\dfrac{13!}{11!3!} =$

20) $\dfrac{9\times210!}{3(7\times30)!} =$

21) $\dfrac{32!}{31!2!} =$

22) $\dfrac{11!12!}{10!13!} =$

23) $\dfrac{16!15!}{14!14!} =$

24) $\dfrac{(5\times3)!}{0!14!} =$

25) $\dfrac{4!(5n-2)!}{(5n)!} =$

26) $\dfrac{4n(4n+7)!}{(4n+8)!} =$

27) $\dfrac{(n-2)!(n+1)}{(n+2)!} =$

Combinations and Permutations

✎ **Calculate the value of each.**

1) 6! = _____

2) 2! × 5! = _____

3) 3 × 4! = _____

4) 5! + 3! = _____

5) 7! = _____

6) 4! = _____

7) 3! + 3! = _____

8) 7! − 5! = _____

✎ **Find the answer for each word problems.**

9) Susan is baking cookies. She uses sugar, butter, Vanilla, eggs and flour. How many different orders of ingredients can she try? _____

10) Albert is planning for his vacation. He wants to go to museum, watch a movie, go to the beach, play the game and play football. How many ways of ordering are there for him? _____

11) How many 4-digit numbers can be named using the digits 3, 4, 5, and 6 without repetition? _____

12) In how many ways can 5 boys be arranged in a straight line? _____

13) In how many ways can 6 athletes be arranged in a straight line? _____

14) A professor is going to arrange her 7 students in a straight line. In how many ways can she do this? _____

15) How many code symbols can be formed with the letters for the word GAMES? _____

16) In how many ways a team of 7 basketball players can choose a captain and co-captain? _____

Answers of Worksheets

Mean and Median

1) Mean: 8.2, Median: 8
2) Mean: 18, Median: 16
3) Mean: 17, Median: 17
4) Mean: 15, Median: 12.5
5) Mean: 16, Median: 17
6) Mean: 29.5, Median: 23
7) Mean: 22, Median: 18
8) Mean: 25, Median: 24
9) Mean: 21, Median: 18
10) Mean: 22.5, Median: 20
11) Mean: 13, Median: 11.5
12) Mean: 13.5, Median: 11
13) Mean: 38.8, Median: 34
14) Mean: 8.8, Median: 8
15) 5

Mode and Range

1) Mode: 3, Range: 4
2) Mode: 18, Range: 18
3) Mode: 8, Range: 14
4) Mode: 24, Range: 14
5) Mode: 27, Range: 21
6) Mode: 8, Range: 15
7) Mode: 13, Range: 13
8) Mode: 5, Range: 15
9) Mode: 7, Range: 14
10) Mode: 15, Range: 15
11) Mode: 6, Range: 40
12) Mode: 8, Range: 30
13) Mode: 12, Range: 45
14) Mode: 15, Range: 8
15) 10

Time series

Day	Distance (km)
1	335
2	496
3	270
4	610
5	320
6	400

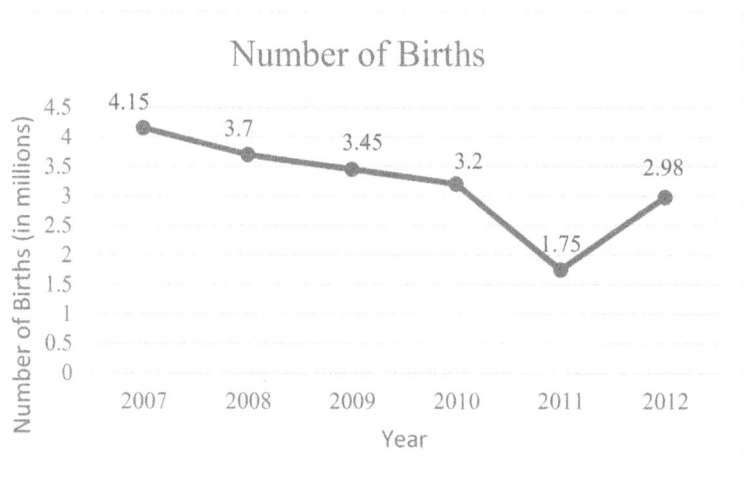

Number of Births

Stem–And–Leaf Plot

1)

Stem	leaf
2	0 1 4 6 7 9
3	0 6 7
5	1 3 5

2)

Stem	leaf
1	1 4 8 9
5	9 9
6	1 5 5 6 8 9

3)

Stem	leaf
5	4 5 9
6	3 6 8
11	2 9
12	1 3 5 8

4)

Stem	leaf
3	2 6 9
5	1 6 6 6
8	4 5 9
10	0 6 7

5)

Stem	leaf
1	1 3 6 7 9
3	0 3 5 5
8	1 6 7 9

6)

Stem	leaf
2	2 1 5 9
6	0 2 4 7
9	2 3 5 8

Pie Graph

1) $5,600

2) $784

3) $1,288

4) $2,016

5) $\frac{1}{2}$

Probability Problems

1) $\frac{3}{5}$

2) $\frac{1}{4}, \frac{2}{3}$

3) $\frac{1}{4}$

4) $\frac{1}{13}$

5) $\frac{7}{15}$

6) $\frac{1}{2}$

7) $\frac{1}{2}$

8) $\frac{1}{8}$

9) 0

Factorials

1) 25

2) 122

3) 4

4) 114

5) 724

6) 9

7) 64

8) 324

9) 26

10) 90

11) 30

12) 6

13) 210

14) $n(n-1)(n-2)$

15) $(n+1)(n+2)$

16) 32

17) $5(n+2)$

18) 19.25

19) 26

20) 3

21) 16

22) $\frac{11}{13}$

23) 3,600

24) 15

25) $\frac{24}{5n(5n-1)}$

26) $\frac{n}{(n+2)}$

27) $\frac{1}{n(n-1)(n+2)}$

Combinations and Permutations

1) 720

2) 240

3) 72

4) 126

5) 5,040

6) 24

7) 12

8) 4,920

9) 120

10) 120

11) 24

12) 120

13) 720

14) 5,040

15) 120

16) 42

Chapter 16 :
ACT Math Test Review

Since 1959, the American College Testing Organization (ACT) has been judging student's potential regarding academics. ACT is a standardized test used for college admissions in the United States. In essence, it is a broad and quick assessment of students' academic abilities. The ACT is divided into four major segments.

- English
- Reading
- Mathematics
- Science

The ACT also includes an optional 40-minute Writing Test.

In an ACT assessment test, all questions are weighted the same. You also have to keep in mind that the more difficult questions are randomly thrown around in the test. You can choose to skip over the more challenging tasks and ace out the simpler questions in the tests first.

There are 60 Mathematics questions on ACT and students have 60 minutes to answer the questions. The Mathematics section of the ACT contains multiple choice questions.

ACT Mathematics cover the following topics:

- Pre-Algebra (20-25%)
- Elementary Algebra (15-20%)
- Intermediate Algebra (15-20%)
- Coordinate Geometry (15-20%)
- Plane Geometry (20-25%)
- Trigonometry (5-10%)

ACT permits the use of personal calculators on the Math portion of the test.

In this section, there are two complete ACT Mathematics Tests. Take these tests to see what score you'll be able to receive on a real ACT test.

Time to Test

Time to refine your skill with a practice examination.

Take a practice ACT Math Test to simulate the test day experience. After you've finished, score your test using the answer key.

Before You Start

- You'll need a pencil, a calculator and a timer to take the test.
- For each question, there are five possible answers. Choose which one is best.
- After you've finished the test, review the answer key to see where you went wrong.

The hardest arithmetic to master is that which enables us to count our blessings.
~Eric Hoffer

Good Luck!

ACT Math Practice Test Answer Sheets

Remove (or photocopy) these answer sheets and use them to complete the practice tests.

ACT Practice Test

1	Ⓐ Ⓑ Ⓒ Ⓓ Ⓔ	21	Ⓐ Ⓑ Ⓒ Ⓓ Ⓔ	41	Ⓐ Ⓑ Ⓒ Ⓓ Ⓔ
2	Ⓐ Ⓑ Ⓒ Ⓓ Ⓔ	22	Ⓐ Ⓑ Ⓒ Ⓓ Ⓔ	42	Ⓐ Ⓑ Ⓒ Ⓓ Ⓔ
3	Ⓐ Ⓑ Ⓒ Ⓓ Ⓔ	23	Ⓐ Ⓑ Ⓒ Ⓓ Ⓔ	43	Ⓐ Ⓑ Ⓒ Ⓓ Ⓔ
4	Ⓐ Ⓑ Ⓒ Ⓓ Ⓔ	24	Ⓐ Ⓑ Ⓒ Ⓓ Ⓔ	44	Ⓐ Ⓑ Ⓒ Ⓓ Ⓔ
5	Ⓐ Ⓑ Ⓒ Ⓓ Ⓔ	25	Ⓐ Ⓑ Ⓒ Ⓓ Ⓔ	45	Ⓐ Ⓑ Ⓒ Ⓓ Ⓔ
6	Ⓐ Ⓑ Ⓒ Ⓓ Ⓔ	26	Ⓐ Ⓑ Ⓒ Ⓓ Ⓔ	46	Ⓐ Ⓑ Ⓒ Ⓓ Ⓔ
7	Ⓐ Ⓑ Ⓒ Ⓓ Ⓔ	27	Ⓐ Ⓑ Ⓒ Ⓓ Ⓔ	47	Ⓐ Ⓑ Ⓒ Ⓓ Ⓔ
8	Ⓐ Ⓑ Ⓒ Ⓓ Ⓔ	28	Ⓐ Ⓑ Ⓒ Ⓓ Ⓔ	48	Ⓐ Ⓑ Ⓒ Ⓓ Ⓔ
9	Ⓐ Ⓑ Ⓒ Ⓓ Ⓔ	29	Ⓐ Ⓑ Ⓒ Ⓓ Ⓔ	49	Ⓐ Ⓑ Ⓒ Ⓓ Ⓔ
10	Ⓐ Ⓑ Ⓒ Ⓓ Ⓔ	30	Ⓐ Ⓑ Ⓒ Ⓓ Ⓔ	50	Ⓐ Ⓑ Ⓒ Ⓓ Ⓔ
11	Ⓐ Ⓑ Ⓒ Ⓓ Ⓔ	31	Ⓐ Ⓑ Ⓒ Ⓓ Ⓔ	51	Ⓐ Ⓑ Ⓒ Ⓓ Ⓔ
12	Ⓐ Ⓑ Ⓒ Ⓓ Ⓔ	32	Ⓐ Ⓑ Ⓒ Ⓓ Ⓔ	52	Ⓐ Ⓑ Ⓒ Ⓓ Ⓔ
13	Ⓐ Ⓑ Ⓒ Ⓓ Ⓔ	33	Ⓐ Ⓑ Ⓒ Ⓓ Ⓔ	53	Ⓐ Ⓑ Ⓒ Ⓓ Ⓔ
14	Ⓐ Ⓑ Ⓒ Ⓓ Ⓔ	34	Ⓐ Ⓑ Ⓒ Ⓓ Ⓔ	54	Ⓐ Ⓑ Ⓒ Ⓓ Ⓔ
15	Ⓐ Ⓑ Ⓒ Ⓓ Ⓔ	35	Ⓐ Ⓑ Ⓒ Ⓓ Ⓔ	55	Ⓐ Ⓑ Ⓒ Ⓓ Ⓔ
16	Ⓐ Ⓑ Ⓒ Ⓓ Ⓔ	36	Ⓐ Ⓑ Ⓒ Ⓓ Ⓔ	56	Ⓐ Ⓑ Ⓒ Ⓓ Ⓔ
17	Ⓐ Ⓑ Ⓒ Ⓓ Ⓔ	37	Ⓐ Ⓑ Ⓒ Ⓓ Ⓔ	57	Ⓐ Ⓑ Ⓒ Ⓓ Ⓔ
18	Ⓐ Ⓑ Ⓒ Ⓓ Ⓔ	38	Ⓐ Ⓑ Ⓒ Ⓓ Ⓔ	58	Ⓐ Ⓑ Ⓒ Ⓓ Ⓔ
19	Ⓐ Ⓑ Ⓒ Ⓓ Ⓔ	39	Ⓐ Ⓑ Ⓒ Ⓓ Ⓔ	59	Ⓐ Ⓑ Ⓒ Ⓓ Ⓔ
20	Ⓐ Ⓑ Ⓒ Ⓓ Ⓔ	40	Ⓐ Ⓑ Ⓒ Ⓓ Ⓔ	60	Ⓐ Ⓑ Ⓒ Ⓓ Ⓔ

ACT Practice Test 1

Mathematics

❖ **60 Questions.**

❖ **Total time for this test: 60 Minutes.**

❖ **You may use a scientific calculator on this test.**

Administered *Month Year*

1) $7^{\frac{8}{3}} \times 7^{\frac{1}{3}} =?$

   A. 7^1 D. 7^4

   B. 7^0 E. 7^2

   C. 7^3

2) If $\frac{6x}{49} = \frac{x-3}{7}$, $x =?$

   A. $\frac{1}{7}$ D. 21

                                  E. $\frac{20}{7}$

   B. $\frac{21}{6}$

   C. 7

3) 115 is equal to?

   A. $34 - (6 \times 7) + (5 \times 16)$

   B. $\left(\frac{14}{9} \times 54\right) + \left(\frac{217}{7}\right)$

   C. $\left(\left(\frac{72}{8} + \frac{24}{4}\right) \times 3\right) - \frac{40}{2} + \frac{160}{4}$

   D. $(4 \times 12) + (38 \times 1.5) - 10$

   E. $\frac{120}{10} + \frac{255}{5}$

4) Six years ago, Amy was four times as old as Mike was. If Mike is 15 years old now, how old is Amy?

   A. 34   D. 30

   B. 48   E. 42

   C. 36

5) A number is chosen at random from 1 to 20. Find the probability of not selecting a composite number.

A. $\frac{7}{20}$ C. $\frac{2}{5}$ E. $\frac{3}{5}$

B. 20   D. 8

6) If $|a| < 5$ then which of the following is true? $(b > 0)$?

   I. $-5b < ba < 5b$

   II. $-a < a^2 < a$ if $a < 0$

   III. $-28 < 4a - 8 < 12$

A. I only   C. I and III only   E. I, II and III

B. II only   D. III only

7) Removing which of the following numbers will change the average of the numbers to 12?

$$7, 10, 11, 14, 17, 18$$

A. 17   C. 11   E. 18

B. 10   D. 7

8) A rope weighs 600 grams per meter of length. What is the weight in kilograms of 21.5 meters of this rope? (1 kilograms = 1,000 grams)

A. 0.0129   C. 12.9   E. 12,900

B. 0.129   D. 1,290

9) If $y = 6ab + 3b^2$, what is y when $a = 3$ and $b = 4$?

A. 46   C. 80   E. 120

B. 60   D. 110

10) If $f(x) = 3 + 4x$ and $g(x) = -4x^2 - 5 - 2x$, then find $(g - f)(x)$?

A. $4x^2 + 6x - 8$ C. $-4x^2 - 6x + 8$ E. $-4x^2 + 6x - 8$

B. $4x^2 - 6x + 8$ D. $-4x^2 - 6x - 8$

11) The marked price of a computer is D dollar. Its price decreased by 60% in January and later increased by 18 % in February. What is the final price of the computer in D dollar?

A. 7.042 D   C. 0.472 D   E. 0.0472 D

B. 47.20 D   D. 472.0 D

12) The number 53.6 is 1,000 times greater than which of the following numbers?

A. 0.536   C. 0.00536   E. 53.60

B. 0.0536   D. 5.360

13) David's current age is 76 years, and Ava's current age is 12 years. In how many years David's age will be 5 times Ava's age?

A. 4   C. 10   E. 16

B. 6   D. 13

14) How many tiles of 12 cm² is needed to cover a floor of dimension 8 cm by 24 cm?

    A. 12   C. 8   E. 24

    B. 36   D. 16

15) What is the median of these numbers? 29, 14, 24, 37, 16, 20, 10

    A. 20   C. 24   E. 18.5

    B. 22   D. 16

16) A company pays its employer $11,000 plus 7% of all sales profit. If x is the number of all sales profit, which of the following represents the employer's revenue?

    A. $0.07x$ C. $0.07x + 11,000$ E. $-0.93x - 11,000$

    B. $0.93x - 11,000$ D. $0.93x + 11,000$

17) What is the area of a square whose diagonal is 4 cm?

    A. 16 cm²   C. 64 cm²   E. 36 cm²

    B. 8 cm²   D. 32 cm²

18) What is the value of x in the following figure?

    A. 75

    B. 150

    C. 45

    D. 135

    E. 145

19) Right triangle ABC is shown below. Which of the following is true for all possible values of angle A and B?

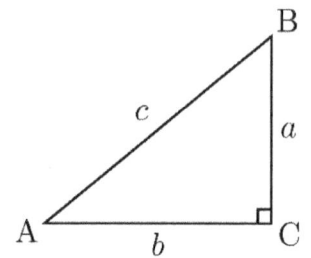

A. $cot\ B\ =\ cos\ A$

B. $tan^2 A = tan^2 B$

C. $tan\ A = 1$

D. $cos\ B\ =\ sin\ A$

E. $\tan A = \sin B$

20) What is the value of y in the following system of equation?

$$3x - 4y = 4$$

$$-x + 3y = 2$$

A. -2

B. -4

C. 2

D. 3

E. 2

21) How long does a 230–miles trip take moving at 50 miles per hour (mph)?

A. 4 hours

B. 4 hours and 6 minutes

C. 6 hours and 36 minutes

D. 4 hours and 40 minutes

E. 6 hours and 46 minutes

22) From the figure, which of the following must be true? (figure not drawn to scale)

A. $y\ =\ 3z$

B. $y\ =\ 4x$

C. $y \geq 2x$

D. $4y + x = z$

E. $y > x$

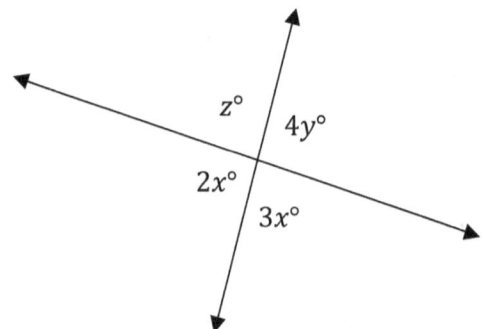

23) Which is the correct statement?

    A. $\frac{1}{4} < 0.18$ D. $\frac{6}{7} > 0.62$

    B. $30\% = \frac{3}{5}$ E. None of them above

    C. $3 < \frac{9}{4}$

24) When 20% of 60 is added to 15% of 400, the resulting number is:

    A. 12   C. 72   E. 124

    B. 60   D. 85

25) A ladder leans against a wall forming a 60° angle between the ground and the ladder. If the bottom of the ladder is 20 feet away from the wall, how long is the ladder?

    A. 20 feet   C. 10 feet   E. 120 feet

    B. 60 feet   D. 40 feet

26) If 65% of a class are girls, and 14% of girls play tennis, what percent of the class play tennis?

    A. 9.1%   C. 0.91%   E. 19%

    B. 0.9%   D. 8%

27) If x is a real number, and if $x^2 + 58 = 110$, then x lies between which two consecutive integers?

    A. 8 and 9   C. 5 and 6   E. 6 and 7

    B. 4 and 5   D. 7 and 8

28) If $(x - 4)^2 = 4$ which of the following could be the value of $(x - 4)(x - 3)$?

A. 4   C. 3   E. −6

B. 6   D. −4

29) Simplify.

$$3x^3 + 6y^5 - x^3 + 4z^3 - 2y^2 + 5x^4 - 3y^5 + 2z^3$$

A. $5x^4 - 2x^3 + 3y^5 + 2y^2 - 6z^3$

B. $5x^4 + 2x^3 + 3y^5 - 2y^2 + 6z^3$

C. $3x^4 + 5x^3 + 2y^5 - 2y^2 + 6z^3$

D. $2x^4 + 3x^3 + 3y^5 - 5y^2 + 6z^3$

E. $3x^4 + 5x^3 + 3y^5 - y^2 + 3z^3$

30) In five successive hours, a car travels 42 km, 39 km, 48 km, 40 km and 50 km. In the next six hours, it travels with an average speed of 52 km per hour. Find the total distance the car traveled in 11 hours.

A. 531 km   C. 219 km   E. 1,351 km

B. 312 km   D. 551 km

31) From last year, the price of gasoline has increased from $1.65 per gallon to $1.98 per gallon. The new price is what percent of the original price?

A. 12%   C. 110%   E. 100%

B. 120%   D. 180%

32) Simplify $(-3 + 4i)(7 + 3i)$.

    A. $33 - 19\,i$ D. $-33 + 19\,i$

    B. $21 - 7i$ E. $7i$

    C. $-40 + 19\,i$

33) If $cot\ \theta = \frac{4}{3}$ and $sin\ \theta > 0$, then $cos\ \theta = ?$

    A. $-\frac{5}{3}$ D. $-\frac{3}{5}$

    B. $\frac{4}{5}$ E. 1

    C. $\frac{5}{4}$

34) Which of the following shows the numbers in increasing order?

    A. $\frac{1}{5}, \frac{8}{9}, \frac{5}{12}, \frac{3}{5}$ D. $\frac{1}{5}, \frac{3}{5}, \frac{5}{12}, \frac{8}{9}$

    B. $\frac{1}{5}, \frac{5}{12}, \frac{3}{5}, \frac{8}{9}$ E. None of them above

    C. $\frac{3}{5}, \frac{8}{9}, \frac{5}{12}, \frac{1}{5}$

35) In 1999, the average worker's income increased \$3,000 per year starting from \$27,000 annual salary. Which equation represents income greater than average? (I = income, x = number of years after 1999)

    A. $I > 3,000\ x + 27,000$ D. $I < 3,000\ x - 27,000$

    B. $I > -3,000\ x + 27,000$ E. $I < 27,000\ x + 3,000$

    C. $I < -3,000\ x + 27,000$

Questions 36 to 38 are based on the following data.

The result of a research shows the number of men and women in four cities of a country.

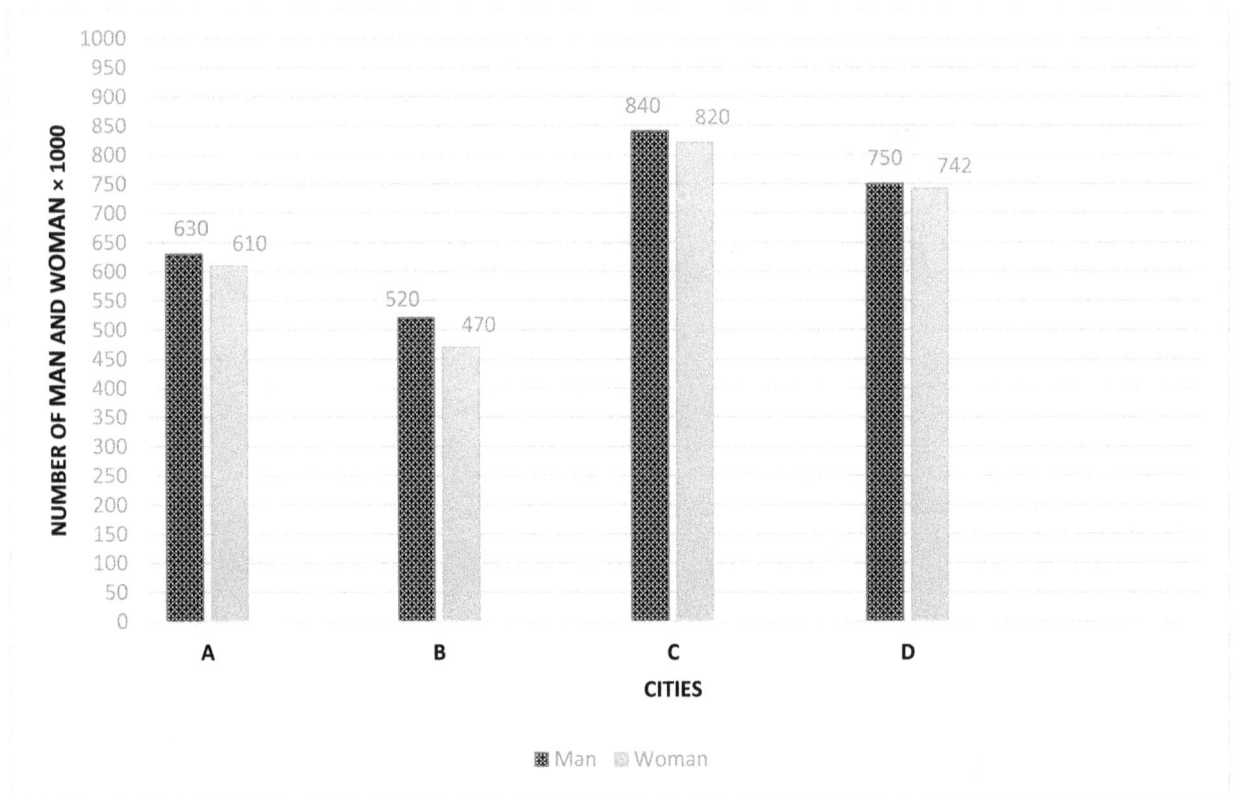

36) What's the ratio of percentage of men in city C to percentage of men in city B?

    A. 0.104   C. 0.96   E. 1.04

    B. 9.633   D. 96.33

37) What's the minimum ratio of woman to man in the four cities?

    A. 0.892   C. 0.904   E. 0.968

    B. 0.989   D. 0.976

38) How many women should be added to city D until the ratio of women to men will be 1.8?

    A. 338   C. 144   E. 334

    B. 608   D. 308

39) What are the values of mode and median in the following set of numbers?

$$7, 3, 9, 6, 9, 9, 7, 7, 7, 9, 3$$

    A. Mode: 7, 9 Median: 9   D. Mode: 3, 9 Median: 3

    B. Mode: 7, 9 Median: 7   E. Mode: 7, Median: 7

    C. Mode: 3, 7 Median: 6

40) y is $x\%$ of what number?

    A. $\dfrac{y}{10x}$ C. $\dfrac{100y}{x}$ E. $\dfrac{xy}{10}$

    B. $\dfrac{100x}{y}$ D. $\dfrac{x}{100y}$

41) If cotangent of an angel β is $\sqrt{3}$, then the tangent of angle β is

    A. $-\sqrt{3}$ C. $\sqrt{3}$ E. $-\dfrac{\sqrt{3}}{3}$

    B. 0   D. $\dfrac{\sqrt{3}}{3}$

42) If a box contains red and blue balls in ratio of 4: 7, how many red balls are there if 84 blue balls are in the box?

    A. 42   C. 32   E. 12

    B. 48   D. 84

ACT Subject Test – Mathematics

43) 9 liters of water are poured into an aquarium that's 30cm long, 5cm wide, and 50cm high. How many cm will the water level in the aquarium rise due to this added water? (1 liter of water = 1,000 cm³)

A. 30

C. 150

E. 15

B. 60

D. 50

44) What is the surface area of the cylinder below?

A. 56 π in²

B. 84 π² in²

C. 48 π in²

D. 36 π² in²

E. 116 π in²

4 in

12

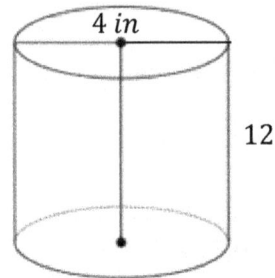

45) A chemical solution contains 7% alcohol. If there is 56ml of alcohol, what is the volume of the solution?

A. 420 ml

C. 800 ml

E. 600 ml

B. 560 ml

D. 1,600 ml

46) What is the solution of the following inequality?

$$|x - 8| \leq 3$$

A. $x \geq 11 \cup x \leq 5$

D. $x \leq 11$

B. $5 \leq x \leq 11$

E. Set of real numbers

C. $x \geq 5$

47) Which of the following points lies on the line $3x - 4y = 13$?

    A. $(2, -1)$ C. $(-1, -4)$ E. $(4, 0)$

    B. $(-2, 2)$ D. $(5, -1)$

48) In the following figure, ABCD is a rectangle, and E and F are points on AD and DC, respectively. The area of ΔBED is 16, and the area of ΔBDF is 15. What is the perimeter of the rectangle?

    A. 28

    B. 14

    C. 40

    D. 24

    E. 38

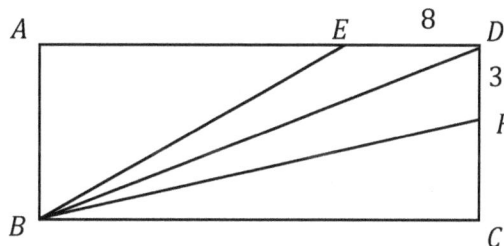

49) In the xy-plane, the point $(2, 7)$ and $(1, 6)$ are on the line A. Which of the following equations of lines is parallel to line A?

    A. $y = -x$ C. $y = \frac{x}{2}$ E. $y = x$

    B. $y = 4$ D. $y = -2x$

50) When point A $(9, 5)$ is reflected over the y-axis to get the point B, what are the coordinates of point B?

    A. $(9, 5)$ C. $(-9, 5)$ E. $(0, 5)$

    B. $(-9, -5)$ D. $(9, -5)$

51) A bag contains 16 balls: two green, five black, four blue, three brown, one red and one white. If 12 balls are removed from the bag at random, what is the probability that a brown ball has been removed?

A. $\frac{1}{4}$

B. $\frac{1}{6}$

C. $\frac{16}{28}$

D. $\frac{3}{16}$

E. $\frac{12}{16}$

52) If a tree casts a 54–foot shadow at the same time that a 4 feet yardstick casts a 9–foot shadow, what is the height of the tree?

A. 14 ft

B. 18 ft

C. 24 ft

D. 28 ft

E. 32 ft

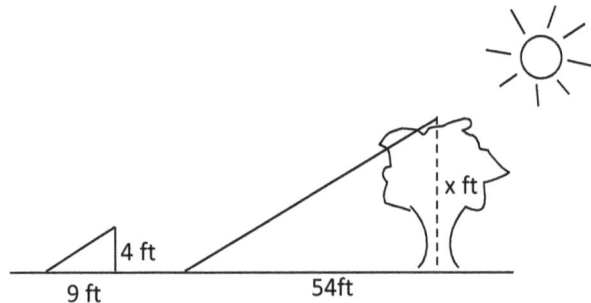

53) If the area of trapezoid is 304, what is the perimeter of the trapezoid?

A. 24

B. 56

C. 10

D. 64

E. 85

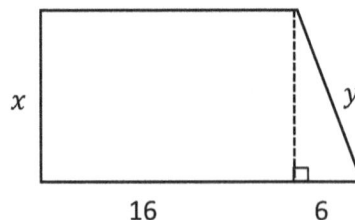

54) If 80% of x equal to 12% of 40, then what is the value of $(x+6)^2$?

A. 480   C. 14.4   E. 144

B. 36   D. 4.8

55) If $f(x) = 4x^4 + 2x^2 - x$ and $g(x) = 2$, what is the value of $f(g(x))$?

A. 62   C. 76   E. 57

B. 28   D. 70

56) A boat sails 24 miles south and then 18 miles east. How far is the boat from its

start point?

A. 25 miles   C. 42 miles   E. 54 miles

B. 30 miles   D. 36 miles

57) If $x \begin{bmatrix} 4 & 0 \\ 0 & 5 \end{bmatrix} = \begin{bmatrix} 3x + 2y - 3 & 0 \\ 0 & 9y - 7 \end{bmatrix}$, what is the product of x and y?

A. 94   C. 72   E. 104

B. 86   D. 96

58) If $f(x) = 3^x$ and $g(x) = log_3 x$, which of the following expressions is equal

to $f(3g(p))$?

A. $3P$ C. p^3 E. $\frac{p}{3}$

B. 3^p D. $3p^3$

59) In the following equation when z is divided by 4, what is the effect on x?

$$x = \frac{6y + \frac{r}{r+4}}{\frac{10}{z}}$$

A. x is divided by 5.

B. x is divided by 4.

C. x does not change.

D. x is multiplied by 4.

E. x is multiplied by 8.

60) Which of the following has the half period and four times the amplitude of graph $y = sinx$?

A. $y = \frac{1}{2}sin\,4x$

B. $y = 4sin\,(\frac{x}{2} + 4)$

C. $y = 2\sin 2x$

D. $y = 4 + 4\,sin\,2x$

E. $y = 2 + sin\,\frac{x}{2}$

STOP

This is the End of this Test. You may check your work on this Test if you still have time.

ACT Practice Test 2

Mathematics

❖ **60 Questions.**

❖ **Total time for this test: 60 Minutes**.

❖ **You may use a scientific calculator on this test.**

Administered *Month Year*

1) Convert 3,260,000 to scientific notation.

    A. $3.26 \times 10{,}000$ D. 3.26×10^6

    B. 3.26×10^{-6} E. 3.26×10^5

    C. $3.26 \times 1{,}000$

2) $(x^4)^{\frac{5}{8}}$ equal to?

    A. $x^{\frac{5}{2}}$ C. $x^{\frac{2}{15}}$ E. $x^{\frac{4}{5}}$

    B. $x^{\frac{2}{5}}$ D. $x^{\frac{5}{4}}$

3) Simplify $\frac{2-5i}{-4i}$?

    A. $\frac{5}{2} + \frac{1}{2}i$ C. $\frac{5}{4} - \frac{1}{2}i$ E. $\frac{5}{4}i$

    B. $\frac{5}{2} - \frac{1}{2}i$ D. $\frac{5}{3} + \frac{5}{3}i$

4) What is the value of x in the following equation?

$$4^x = 256$$

    A. 5   C. 7   E. 3

    B. 4   D. 6

5) What is the sum of prime numbers between 1 and 20?

    A. 88   C. 75   E. 78

    B. 87   D. 77

6) If $\sqrt[3]{6x} = \sqrt[3]{y}$, then $x =$?

   A. $6y$ C. y^3 E. $\dfrac{y}{6}$

   B. $\sqrt[3]{\dfrac{y}{6}}$ D. $\sqrt{6y}$

7) The average weight of 18 girls in a class is 45 kg and the average weight of 22 boys in the same class is 50 kg. What is the average weight of all the 40 students in that class?

   A. 49.25   C. 49.50   E. 47.25

   B. 47.75   D. 49.75

8) If $y = (-3x^6)^3$, which of the following expressions is equal to y?

   A. $-9x^9$ C. $9x^9$ E. $-27x^{18}$

   B. $-81x^9$ D. $27x^{18}$

9) What is the value of the expression $5(x + y) + (7 - x)^3$ when $x = 5$ and $y = -3$?

   A. -15 C. 18   E. 15

   B. -2 D. -18

10) Sophia purchased a sofa for $246.50 The sofa is regularly priced at $725. What was the percent discount Sophia received on the sofa?

   A. 1.66%   C. 43%   E. 1.34%

   B. 66%   D. 34%

11) If $f(x) = 2x - 9$ and $g(x) = 4x^2 - 3x$, then find $(\frac{f}{g})(x)$.

A. $\dfrac{2x-9}{4x^2-3x}$ C. $\dfrac{x-4}{x^2-3}$ E. $\dfrac{x^2-3x}{2x-9}$

B. $\dfrac{x-9}{4x^2-3x}$ D. $\dfrac{2x+9}{x^2+4x}$

12) In the standard (x, y) coordinate plane, which of the following lines contains

the points $(2, -1)$ and $(4, 7)$?

A. $y = 4x - 9$

B. $y = \dfrac{1}{4}x + 9$

C. $y = -4x + 9$

D. $y = -\dfrac{1}{4}x + 9$

E. $y = 4x - 9$

13) A bank is offering 2.25% simple interest on a savings account. If you deposit

$24,000, how much interest will you earn in four years?

A. $2,400   C. $1,620   E. $2,600

B. $2,160   D. $2,610

14) If the ratio of home fans to visiting fans in a crowd is 2:7and all 18,000 seats in

a stadium are filled, how many visiting fans are in attendance?

A. 162,000   C. 4,000   E. 14,000

B. 126,000   D. 36,000

15) If the interior angles of a quadrilateral are in the ratio 3:7:9:11, what is the measure of the largest angle?

A. 108°

C. 152°

E. 132°

B. 84°

D. 162°

16) If $x + 4sin^2a + 4cos^2a = 10$, then $x = ?$

A. 1

C. 14

E. 2

B. 6

D. 4

17) If the area of a circle is 121 square meters, what is its diameter?

A. 11π

C. $\frac{22\sqrt{\pi}}{\pi}$

E. $22\sqrt{\pi}$

B. $\frac{11}{\pi}$

D. $121\pi^2$

18) The length of a rectangle is $\frac{5}{7}$ times its width. If the width is 70, what is the perimeter of this rectangle?

A. 700

C. 240

E. 50

B. 350

D. 120

19) In the figure below, line A is parallel to line B. What is the value of angle x?

A. 65 degree

B. 125 degree

C. 105 degree

D. 135 degree

E. 115 degree

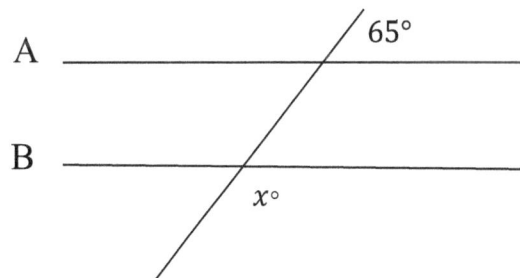

20) What is the value of x in the following system of equations?

$$x + 3y = 14$$
$$3x + 2y = 21$$

A. 5   C. -5 E. 8

B. 3   D. -3

21) An angle is equal to one eleventh of its supplement. What is the measure of that

angle?

A. 15   C. 20   E. 18

B. 16   D. 12

22) Last week 12,000 fans attended a football match. This week six times as many

bought tickets, but one ninth of them cancelled their tickets. How many are

attending this week?

A. 8,000   C. 46,000   E. 32,000

B. 4,000   D. 64,000

23) If $sin\alpha = \frac{\sqrt{2}}{2}$ in a right triangle and the angle α is an acute angle, then what is

$cos\ \alpha$?

A. $\frac{1}{2}$ C. $\sqrt{2}$ E. $\frac{\sqrt{2}}{2}$

B. 1   D. 0

24) In following **squares** which statement is true?

    A. AB is parallel to BC.

    B. AB is perpendicular to DC.

    C. Length of AB equal to half of length BC.

    D. The measure of all the angles equals $360°$.

    E. The answer cannot be found from the information given.

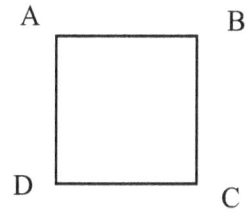

25) In the standard (x, y) coordinate system plane, what is the area of the circle with the following equation?

$$(x + 4)^2 + (y - 3)^2 = 16$$

    A. 4π C. 8π E. 2π

    B. 16π D. 32π

26) In two successive years, the population of a town is increased by 18% and 20%. What percent of the population is increased after two years?

    A. 40%   C. 41.6%   E. 46.5%

    B. 46%   D. 36.6%

27) Simplify.

$$5x^6y^2 + 4x^3y^5 - (3x^6y^2 - 6x^3y^5)$$

    A. $-10x^6y^3$ C. $2x^6y^2$ E. $12x^5y^6$

    B. $2x^6y^2 - 10x^3y^5$ D. $2x^6y^2 + 10x^3y^5$

28) What are the zeroes of the function $f(x) = 4x^3 + 40x^2 + 84x$?

A. 0

C. 0, 3, 7

E. 0, −3, −7

B. −3, 7

D. −7, −3

29) If one angle of a right triangle measures 60°, what is the sine of the other acute angle?

A. $\frac{1}{2}$

C. $\frac{\sqrt{3}}{2}$

E. $\sqrt{3}$

B. $\frac{\sqrt{3}}{3}$

D. 0

30) In the following figure, what is the perimeter of ΔABC if the area of ΔADC is 48?

A. 32

B. 24

C. 84

D. 48

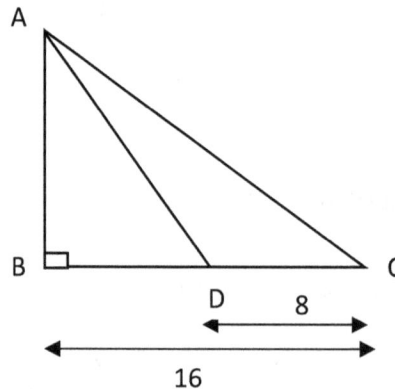

E. The answer cannot be determined from the information given.

31) Which of the following is one solution of this equation?

$$4x^2 + 7x - 2 = 0$$

A. $\sqrt{3} + 1$

C. −2

E. $\sqrt{2}$

B. $\sqrt{3} - 1$

D. $\sqrt{3}$

32) Four-kilograms apple and five-kilograms orange cost $72.8. If one-kilogram apple costs $3.2 how much does one-kilogram orange cost?

    A. $12   C. $14.5   E. $14

    B. $10   D. $6

33) Which of the following expressions is equal to $\sqrt{\frac{x^2}{7} + \frac{2x^2}{49}}$?

    A. $3x$ C. $3x\sqrt{x}$ E. $7x$

    B. $\frac{3x}{7}$ D. $\frac{x\sqrt{x}}{7}$

34) If the ratio of $16a$ to $12b$ is $\frac{1}{20}$, what is the ratio of a to b?

    A. 3   C. $\frac{3}{80}$ E. $\frac{1}{40}$

    B. 80   D. $\frac{3}{40}$

35) In the following figure, ABCD is a rectangle. If $a = 2\sqrt{2}$, and $b = 2a$, find the area of the shaded region. (the shaded region is a trapezoid)

    A. 20

    B. 12

    C. $15\sqrt{2}$

    D. $14\sqrt{2}$

    E. $16\sqrt{2}$

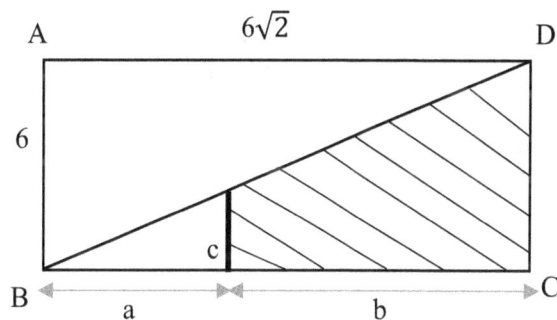

Questions 36 to 38 are based on the following data.

Number of clothes sold in a clothing store

■ Shirts   ■ Pants   ▥ Shoes

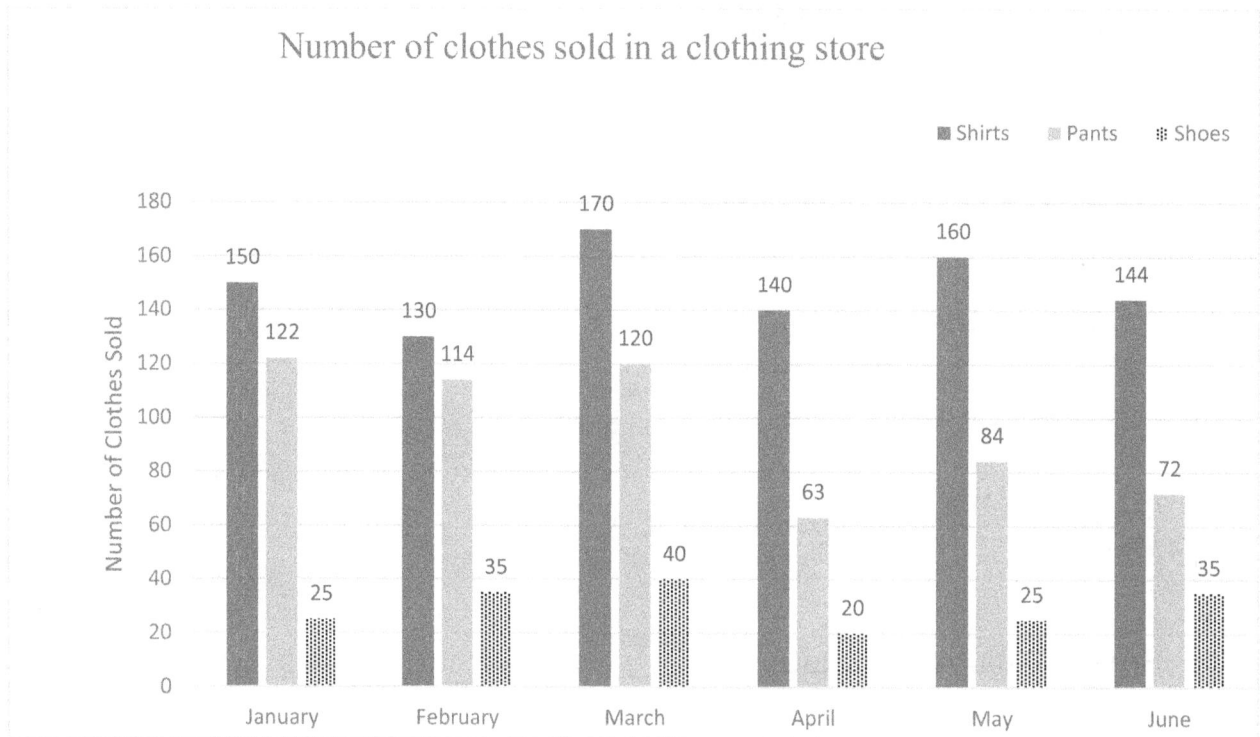

36) Between which two of the months shown was there a twenty-five percent increase in the number of pants sold?

A. January and February   D. April and May

B. February and March   E. May and June

C. March and April

37) During the six-month period shown, what is the mean number of shirts and median number of shoes per month?

A. 147.5, 40   D. 145, 35

B. 139, 45   E. 194, 60

C. 149, 30

38) How many shoes need to be added in February until the ratio of number of pants to number of shoes in February equals to three-twelfth of this ratio in March?

A. 120   C. 115   E. 70

B. 117   D. 118

39) A card is drawn at random from a standard 52–card deck, what is the probability that the card is of spades or diamonds? (The deck includes 13 of each suit clubs, diamonds, hearts, and spades)

A. $\frac{1}{4}$ C. $\frac{2}{13}$ E. $\frac{1}{26}$

B. $\frac{1}{2}$ D. $\frac{1}{54}$

40) A football team had $34,000 to spend on supplies. The team spent $21,000 on new balls. New sport shoes cost $120 each. Which of the following inequalities represent how many new shoes the team can purchase?

A. $120x + 21,000 \leq 34,000$ D. $21,000x + 120 \geq 34,000$

B. $120x + 21,000 \geq 34,000$ E. $21,000x + 34,000 \geq 120$

C. $34,000x + 120 \leq 21,000$

41) If $x = 2$, what is the value of y in the following equation? $3y = \frac{7x^4}{16} + 14$

A. 7   C. 21   E. 126

B. 14   D. 3

42) A swimming pool holds 6,000 cubic feet of water. The swimming pool is 20 feet long and 6 feet wide. How deep is the swimming pool?

A. 300 feet
C. 60 feet
E. 50 feet

B. 120 feet
D. 40 feet

43) The ratio of boys to girls in a school is 9:11. If there are 480 students in a school, how many boys are in the school.

A. 117
C. 195
E. 216

B. 252
D. 264

44) If $(x - 5)^2 + 2 > 4x + 5$, then x can equal which of the following?

A. 2
C. 0
E. 4

B. 6
D. 5

45) Let r and p be constants. If $x^2 + 7x + r$ factors into $(x + 5)(x + p)$, the values of r and p respectively are?

A. 10, 2
C. 7, 5

B. 5, 4
D. 7, 2

E. The answer cannot be found from the information given.

46) If 160% of a number is 72, then what is 40% of that number?

A. 54
C. 18
E. 45

B. 22
D. 81

47) The width of a box is third of its length. The height of the box is third of its width. If the length of the box is 45 cm, what is the volume of the box?

A. 675 cm³

B. 2,700 cm³

C. 3,537 cm³

D. 3,375 cm³

E. 3,735 cm³

48) The average of six consecutive numbers is 35. What is the smallest number?

A. 22.5

B. 33.5

C. 32.5

D. 35.5

E. 37.5

49) The surface area of a cylinder is $112\pi\ cm^2$. If its height is 10 cm, what is the radius of the cylinder?

A. 5 cm

B. 15 cm

C. 56 cm

D. 4 cm

E. 14 cm

50) In a coordinate plane, triangle ABC has coordinates: $(7, -2)$, $(-5, -3)$, and $(4, 6)$. If triangle ABC is reflected over the y-axis, what are the coordinates of the new image?

A. $(7, -2), (-3, -5), (6, 4)$

B. $(-2, 7), (-3, -5), (6, 4)$

C. $(-7, -2), (5, -3), (-4, 6)$

D. $(7, -2), (3, 5), (4, -6)$

E. $(-7, -2), (-5, 3), (4, 6)$

51) What is the slope of a line that is perpendicular to the line $14x - 2y = 8$?

   A. -7

   B. $-\frac{1}{7}$

   C. 7

   D. -4

   E. $\frac{1}{7}$

52) What is the difference in area between a 7 cm by 5 cm rectangle and a circle with diameter of 10 cm? ($\pi = 3$)

   A. 50

   B. 40

   C. 30

   D. 110

   E. 35

53) If $f(x)=5x^2+2$ and $g(x)=\frac{3}{x}$, what is the value of $f(g(x))$?

   A. $\frac{3}{5x^2+2}$

   B. $\frac{5}{x^2}$

   C. $\frac{5}{3x}$

   D. $\frac{1}{5x+2}$

   E. $\frac{45}{x^2}+2$

54) A cruise line ship left Port A and traveled 18 miles due west and then 24 miles due north. At this point, what is the shortest distance from the cruise to port A?

   A. 49 miles

   B. 42 miles

   C. 32 miles

   D. 30 miles

   E. 36 miles

55) The length of a rectangle is 3 meters greater than 5 times its width. The perimeter of the rectangle is 114 meters. What is the area of the rectangle?

   A. $57 \, \text{m}^2$

   B. $513 \, \text{m}^2$

   C. $432 \, \text{m}^2$

   D. $384 \, \text{m}^2$

   E. $423 \, \text{m}^2$

ACT Subject Test – Mathematics

56) Tickets to a movie cost $9.50 for adults and $6.50 for students. A group of 22 friends purchased tickets for $167. How many student tickets did they buy?

A. 12   C. 6   E. 18

B. 14   D. 8

57) What is the solution of the following inequality?

$$|x - 6| \geq 10$$

A. $x \geq 16 \cup x \leq -4$ D. $x \leq -4$

B. $-4 \leq x \leq 16$ E. Set of real numbers

C. $x \geq 16$

58) If $\tan x = \frac{10}{24}$, then $\sin x =$

A. $\frac{1}{12}$ C. $\frac{24}{26}$

B. $\frac{5}{13}$ D. $\frac{7}{13}$

E. It cannot be determined from the information given.

59) If $A = \begin{bmatrix} 2 & 3 \\ -1 & 4 \end{bmatrix}$ and $B = \begin{bmatrix} 5 & 1 \\ -1 & 4 \end{bmatrix}$, then $2A - B =$

A. $\begin{bmatrix} 1 & 4 \\ 1 & -5 \end{bmatrix}$ C. $\begin{bmatrix} 9 & 7 \\ -3 & 12 \end{bmatrix}$ E. $\begin{bmatrix} -1 & 5 \\ -1 & 4 \end{bmatrix}$

B. $\begin{bmatrix} -5 & 4 \\ 1 & -2 \end{bmatrix}$ D. $\begin{bmatrix} -1 & 4 \\ -1 & 5 \end{bmatrix}$

60) What is the amplitude of the graph of the equation $y - 3 = 6cos4x$?

(half the distance between the graph's minimum and maximum y-values in standard (x, y) coordinate plane is the amplitude of a graph.)

A. 3

C. 0.3

E. 0.6

B. 6

D. 4

STOP

This is the End of this Test. You may check your work on this Test if you still have time.

Chapter 17 :
Answers and Explanations

Answer Key

❋ Now, it's time to review your results to see where you went wrong and what areas you need to improve!

ACT Math Practice Test

	Practice Test 1							Practice Test 2					
1	C	21	C	41	D		1	D	21	A	41	A
2	D	22	D	42	B		2	A	22	D	42	E
3	B	23	D	43	B		3	A	23	E	43	E
4	E	24	C	44	A		4	B	24	D	44	C
5	C	25	D	45	C		5	D	25	B	45	A
6	C	26	A	46	B		6	E	26	C	46	C
7	A	27	D	47	C		7	B	27	D	47	D
8	C	28	B	48	A		8	E	28	E	48	C
9	E	29	B	49	E		9	C	29	A	49	D
10	D	30	A	50	C		10	B	30	D	50	C
11	B	31	B	51	D		11	A	31	C	51	B
12	B	32	D	52	C		12	A	32	A	52	B
13	A	33	B	53	B		13	B	33	B	53	E
14	D	34	B	54	E		14	E	34	C	54	D
15	A	35	A	55	D		15	E	35	E	55	C
16	B	36	D	56	B		16	B	36	D	56	B
17	C	37	C	57	E		17	C	37	C	57	A
18	B	38	B	58	C		18	C	38	B	58	B
19	D	39	B	59	B		19	E	39	B	59	E
20	E	40	C	60	D		20	A	40	A	60	B

ACT Mathematics

Practice Tests 1

1) Answer: C.

$7^{\frac{8}{3}} \times 7^{\frac{1}{3}} = 7^{\frac{8}{3}+\frac{1}{3}} = 7^{\frac{9}{3}} = 7^3$

2) Answer: D.

Solve for x, $\frac{6x}{49} = \frac{x-3}{7}$

Multiply the second fraction by 7, $\frac{6x}{49} = \frac{7(x-3)}{7 \times 7}$

Tow denominators are equal. Therefore, the numerators must be equal.

$6x = 7x - 21 \rightarrow -x = -21 \rightarrow x = 21$

3) Answer: B.

Simplify each option provided.

A. $34 - (6 \times 7) + (5 \times 16) = 34 - 42 + 80 = 72$

B. $\left(\frac{14}{9} \times 54\right) + \left(\frac{217}{7}\right) = 84 + 31 = 115$ (this is the answer)

C. $\left(\left(\frac{72}{8} + \frac{24}{4}\right) \times 3\right) - \frac{40}{2} + \frac{160}{4} = \left(\left(\frac{72+48}{8}\right) \times 3\right) - \frac{40}{2} + \frac{160}{4} = \left(\left(\frac{120}{8}\right) \times 3\right) +$

$\frac{160-80}{4} = (15 \times 3) + \frac{80}{4} = 45 + 20 = 65$

D. $(4 \times 12) + (48 \times 1.5) - 10 = 48 + 72 - 10 = 110$

E. $\frac{120}{10} + \frac{255}{5} = \frac{120+510}{10} = 63$

4) Answer: E.

six years ago, Amy was four times as old as Mike. Mike is 15 years.

now. Therefore, 6 years ago Mike was 9 years.

six years ago, Amy was: $A = 4 \times 9 = 36$

Now Amy is 42 years old: $36 + 6 = 42$

5) Answer: C.

Set of number that are not composite between 1 and 20: A = {2, 3, 5, 7, 11, 13, 17, 19}

Probability $= \frac{number\ of\ desired\ outcomes}{number\ of\ total\ outcomes} = \frac{8}{20} = \frac{2}{5}$

6) Answer: C.

I. $|a| < 5 \rightarrow -5 < a < 5$

Multiply all sides by b. Since $b > 0 \rightarrow -5b < ba < 5b$ (it is true!)

II. Since, $-5 < a < 5, and\ a < 0 \rightarrow -a > a^2 > a$ (plug in $-\frac{1}{5}$, and check!) (It's false)

III. $-5 < a < 5, multiply\ all\ sides\ by\ 4, then: -20 < 4a < 20$

Subtract 8 from all sides. Then:

$-20 - 8 < 4a - 8 < 20 - 8 \rightarrow -28 < 4a - 8 < 12$ (It is true!)

7) Answer: A.

Check each option provided:

A. 17 $\frac{7+10+11+14+18}{5} = \frac{60}{5} = 12$

B. 15 $\frac{7+11+14+17+18}{5} = \frac{67}{5} = 13.4$

C. 11 $\frac{7+10+14+17+18}{5} = \frac{66}{5} = 13.2$

D. 7 $\frac{10+11+17+14+18}{5} = \frac{70}{5} = 14$

E. 18 $\frac{7+10+11+14+17}{5} = \frac{59}{5} = 11.8$

8) Answer: C.

The weight of 21.5 meters of this rope is: $21.5 \times 600\ g = 12,900\ g$

1 kg = 1,000 g, therefore, 12,900 g ÷ 1,000 = 12.9 kg

9) Answer: E.

$y = 6ab + 3b^2$

Plug in the values of a and b in the equation: $a = 3$ and $b = 4$

$y = 6\ (3)\ (4) + 3\ (4)^2 = 72 + 3(16) = 72 + 48 = 120$

10) Answer: D.

$(g- f)(x) = g(x) - f(x) = (-4x^2 - 5 - 2x) - (3 + 4x)$

$-4x^2 - 5 - 2x - 3 - 4x = -4x^2 - 6x - 8$

11) Answer: B.

To find the discount, multiply the number by (100% – rate of discount).

Therefore, for the first discount we get: (D) (100% – 60%) = (D) (0.40) = 0.40 D

For increase of 18 %: (0.40 D) (100% + 18%) = (0.40 D) (1.18) = 0.472 D = 47.20%

of D

12) Answer: B.

1,000 times the number is 53.6. Let x be the number, then:

$1,000x = 53.6 \rightarrow x = \frac{53.6}{1,000} = 0.0536$

13) Answer: A.

Let's review the options provided.

A. 4. In 4 years, David will be 80 and Ava will be 16. 80 is 5 times 16.

B. 6. In 6 years, David will be 82 and Ava will be 18. 82 is NOT 5 times 18.

C. 10. In 10 years, David will be 86 and Ava will be 22. 86 is not 5 times 22.

D. 13. In 13 years, David will be 89 and Ava will be 25. 89 is not 5 times 25.

E. 16. In 16 years, David will be 92 and Ava will be 28. 92 is not 5 times 28.

14) Answer: D.

The area of the floor is: 8 cm × 24cm = 192 cm

The number is tiles needed = 192 ÷ 12 = 16

15) Answer: A.

Write the numbers in order: 10, 14, 16, 20, 24, 29, 37.

Since we have 7 numbers (7 is odd), then the median is the number in the middle,

which is 20.

16) Answer: B.

The diagonal of the square is 4. Let x be the side.

Use Pythagorean Theorem: $a^2 + b^2 = c^2$

$x^2 + x^2 = 4^2 \Rightarrow 2x^2 = 16 \Rightarrow 2x^2 = 16 \Rightarrow x^2 = 8 \Rightarrow x = \sqrt{8}$

The area of the square is: $\sqrt{8} \times \sqrt{8} = 8$

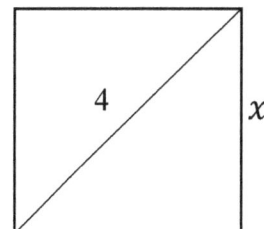

17) Answer: C.

Employer's revenue: $0.07x + 11,000$

18) Answer: B.

$x = 45 + 105 = 150$

19) Answer: D.

By definition, the sine of any acute angle is equal to the cosine of its complement.

Since, angle A and B are complementary angles, therefore:

$cos B = sin A$

20) Answer: E.

Solve the system of equations by elimination method.

$3x - 4y = 4$
$-x + 3y = 2$ Multiply the second equation by 3, then add it to the first equation.

$\begin{matrix} 3x - 4y = 4 \\ 3(-x + 3y = 2) \end{matrix} \Rightarrow \begin{matrix} 3x - 4y = 4 \\ -3x + 9y = 6 \end{matrix} \Rightarrow$ add the equations, $y = 2$

21) Answer: C.

Use distance formula:

Distance = Rate × time $\Rightarrow 230 = 50 \times T$, divide both sides by 50.

$\frac{230}{50} = T \Rightarrow T = 4.6$ hours.

Change hours to minutes for the decimal part. 0.6 hours $= 0.6 \times 60 = 36$ minutes.

22) Answer: D.

$2x$ and z are colinear. $4y$ and $3x$ are colinear. Therefore,

$2x + z = 4y + 3x, subtract\ 2x\ from\ both\ sides,, then, z = 4y + x$

23) Answer: D.

Check each option.

$\frac{1}{4} < 0.18$ $\frac{1}{4} = 0.25$ and it is less than 0.18. Not true!

A. $30\% = \frac{3}{5}$ $30\% = \frac{3}{10} < \frac{3}{5} = \frac{6}{10}$. Not True!

B. $3 < \frac{9}{4}$ $\frac{9}{4} = 2.25 < 4$. Not True!

C. $\frac{6}{7} > 0.62$ $\frac{6}{7} = 0.857$ and it is greater than 0.62. Bingo!

D. None of them above   Not True!

24) Answer: C.

20% of 60 equals to: $0.20 \times 60 = 12$

15% of 400 equals to: $0.15 \times 400 = 60$

20% of 60 is added to 15% of 400: $12 + 60 = 72$

25) Answer: D.

The relationship among all sides of special right triangle

$30°, \ 60°, \ 90°$ is provided in this triangle:

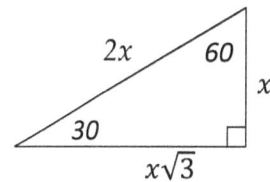

In this triangle, the opposite side of 30° angle is half of the

hypotenuse. Draw the shape of this question.

The ladder is the hypotenuse.

Therefore, the ladder is 40 ft.

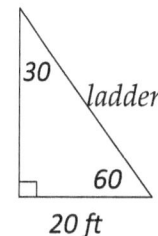

26) Answer: A.

The percent of girls playing tennis is: $65 \ \% \times 14 \ \% = 0.65 \times 0.14 = 0.091 = 9.1\%$

27) Answer: D.

Solve for x. $x^2 + 58 = 110 \Rightarrow x^2 = 52$

Let's review the options.

A. 8 and 9. $8^2 = 64$ and $9^2 = 81$, 52 is not between these two numbers.

B. 4 and 5. $4^2 = 16$ and $5^2 = 25$, 52 is not between these two numbers.

C. 5 and 6. $5^2 = 25$ and $5^2 = 36$, 52 is not between these two numbers.

D. 7 and 8. $7^2 = 49$ and $8^2 = 64$, 52 is between these two numbers.

E. 6 and 7. $6^2 = 36$ and $7^2 = 49$, 57 is not between these two numbers.

28) Answer: B.

$(x - 4)^2 = 4 \to x - 4 = 2 \to x = 6$

$\to (x - 4)(x - 3) = (6 - 4)(6 - 3) = (2)(3) = 6$

29) Answer: B.

$3x^3 + 6y^5 - x^3 + 4z^3 - 2y^2 + 5x^4 - 3y^5 + 2z^3 = 3x^3 - x^3 + 5x^4 + 6y^5 - 3y^5 - 2y^2 + 4z^3 + 2z^3 = 5x^4 + 2x^3 + 3y^5 - 2y^2 + 6z^3$

30) Answer: A.

Add the first 5 numbers. $42 + 39 + 48 + 40 + 50 = 219$

To find the distance traveled in the next 6 hours, multiply the average by number of hours.

Distance = Average × Rate = $52 \times 6 = 312$

Add both numbers. $219 + 312 = 531$

31) Answer: B.

The question is this: 1.98 is what percent of 1.65?

Use percent formula: part = $\frac{percent}{100}$ × whole $\Rightarrow 1.98 = \frac{percent}{100} \times 1.65$

$\Rightarrow 1.98 = \frac{percent \times 1.65}{100} \Rightarrow 198 = percent \times 1.65 \Rightarrow percent = \frac{198}{1.65} = 120$

32) Answer: D.

We know that: $i = \sqrt{-1} \Rightarrow i^2 = -1$

$(-3 + 4i)(7 + 3i) = -21 - 9i + 28i + 12i^2 = -21 + 19i - 12 = 19i - 33$

33) Answer: B.

$cot\theta = \frac{adjacent}{opposite}$

$cot\theta = \frac{4}{3} \Rightarrow$ we have the following right triangle. Then,

$c = \sqrt{3^2 + 4^2} = \sqrt{9 + 16} = \sqrt{25} = 5$

$cos\theta = \frac{adjacent}{hypotenuse} = \frac{4}{5}$

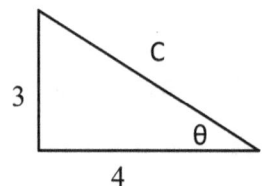

34) Answer: B.

$\frac{1}{5} = 0.2 \qquad \frac{5}{12} \cong 0.42 \qquad \frac{3}{5} = 0.6 \qquad \frac{8}{9} \cong 0.89$

35) Answer: A.

Let x be the number of years. Therefore, $3,000 per year equals $3,000x$.

starting from $27,000 annual salary means you should add that amount to $3,000x$.

Income more than that is: $I > 3,000\,x + 27,000$.

36) Answer: D.

Percentage of men in city C $= \dfrac{840}{1,660} \times 100 = 50.60\%$

Percentage of men in city B $= \dfrac{520}{990} \times 100 = 52.53\%$

Percentage of men in city C to percentage of men in city B: $\dfrac{50.60}{52.53} = 96.33$

37) Answer: C.

Ratio of women to men in city A: $\dfrac{610}{630} = 0.968$

Ratio of women to men in city B: $\dfrac{470}{520} = 0.904$

Ratio of women to men in city C: $\dfrac{820}{840} = 0.976$

Ratio of women to men in city D: $\dfrac{742}{750} = 0.989$

38) Answer: B.

Let the number of women should be added to city D be x, then:

$\dfrac{742 + x}{750} = 1.8 \rightarrow 742 + x = 750 \times 1.8 = 1,350 \rightarrow x = 608$

39) Answer: B.

We write the numbers in the order: 3, 3, 6, 7, 7, 7, 7, 9, 9, 9, 9.

The mode of numbers is: 7 and 9, median is: 7.

40) Answer: C.

Let the number be A. Then: $y = x\% \times A \rightarrow$ (Solve for A) $\rightarrow x = \dfrac{x}{100} \times A$

Multiply both sides by $\dfrac{100}{x}$: $y \times \dfrac{100}{x} = \dfrac{x}{100} \times \dfrac{100}{x} \times A \rightarrow A = \dfrac{100y}{x}$

41) Answer: D.

$tangent\ \beta = \dfrac{1}{cotangent\ \beta} = \dfrac{1}{\sqrt{3}} = \dfrac{\sqrt{3}}{3}$

42) Answer: B.

$\dfrac{4}{7} \times 84 = 48$

43) Answer: B.

One liter=$1,000 cm^3 \rightarrow$ 9 liters = $9,000\ cm^3$

$9,000 = 30 \times 5 \times h \rightarrow h = \frac{9,000}{150} = 60$ cm

44) Answer: A.

Surface Area of a cylinder = $2\pi r\ (r + h)$,

The radius of the cylinder is 2 ($4 \div 2$) inches, and its height is 12 inches. Therefore,

Surface Area of a cylinder = $2\pi\ (2)\ (2 + 12) = 56\ \pi$

45) Answer: C.

7% of the volume of the solution is alcohol. Let x be the volume of the solution.

Then: 7% of $x = 56$ml $\Rightarrow 0.07\ x = 56 \Rightarrow x = 56 \div 0.07 = 800$

46) Answer: B.

$|x - 8| \leq 3 \rightarrow -3 \leq x - 8 \leq 3 \rightarrow -3 + 8 \leq x - 8 + 8 \leq 3 + 8 \rightarrow 5 \leq x \leq 11$

47) Answer: C.

Plug in each pair of number in the equation:

A. $(2, -1)$: $3(2) - 4(-1) = 10$ Nope!

B. $(-2, 2)$: $3(-2) - 4(2) = -14$ Nope!

C. $(-1, -4)$: $3(-1) - 4(-4) = 13$ Bingo!

D. $(5, -1)$: $3(5) - 4(-1) = 19$ Nope!

$(4, 0)$: $3(4) - 4(0) = 12$ Nope!

48) Answer: A.

The area of ΔBED is 16, then: $\frac{8 \times AB}{2} = 16 \rightarrow 8 \times AB = 32 \rightarrow AB = 4$

The area of ΔBDF is 15, then: $\frac{3 \times BC}{2} = 15 \rightarrow 3 \times BC = 30 \rightarrow BC = 10$

The perimeter of the rectangle is = $2 \times (4 + 10) = 28$

49) Answer: E.

The slop of line A is: m $= \frac{y_2 - y_1}{x_2 - x_1} = \frac{7 - 6}{2 - 1} = 1$

Parallel lines have the same slope and only choice E ($y = x$) has slope of 1.

50) Answer: C.

When points are reflected over y-axis, the value of y in the coordinates doesn't change and the sign of x changes. Therefore, the coordinates of point B are $(-9, 5)$.

51) Answer: D.

If 16 balls are removed from the bag at random, there will be three ball in the bag. The probability of choosing three brown ball is 3 out of 12. Therefore, the probability of not choosing a brown ball is 12 out of 16 and the probability of having not a brown ball after removing 12 balls is the same.

52) Answer: C.

Write a proportion and solve for x.

$\frac{4}{9} = \frac{x}{54} \Rightarrow 9x = 4 \times 54 \Rightarrow x = 24$ ft

53) Answer: B.

The area of trapezoid is: $\left(\frac{16+22}{2}\right) \times x = 304 \rightarrow 38x = 304 \rightarrow x = 8$

$y = \sqrt{6^2 + 8^2} = 10$

Perimeter is: $22 + 8 + 16 + 10 = 56$

54) Answer: E.

$0.8x = (0.12) \times 40 \rightarrow x = 6 \rightarrow (x+6)^2 = (12)^2 = 144$

55) Answer: D.

$g(x) = 2$,

then $f(g(x)) = f(2) = 4(2)^4 + 2(2)^2 - (2) = 64 + 8 - 2 = 70$

56) Answer: B.

Use the information provided in the question to draw the shape.

Use Pythagorean Theorem: $a^2 + b^2 = c^2$

$18^2 + 24^2 = c^2 \Rightarrow 324 + 576 = c^2$

$\Rightarrow 900 = c^2 \Rightarrow c = 30$

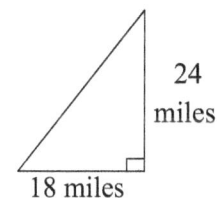

57) Answer: E.

$\begin{cases} 4x = 3x + 2y - 3 \\ 5x = 9y - 7 \end{cases} \rightarrow \begin{cases} x - 2y = -3 \\ 5x - 9y = -7 \end{cases}$

Multiply first equation by -5.

$$\begin{cases} -5x + 10y = 15 \\ 5x - 9y = -7 \end{cases} \rightarrow \text{add two equations.}$$

$y = 8 \rightarrow y = 8 \rightarrow x = 13 \rightarrow x \times y = 104$

58) Answer: C.

To solve for $f(3g(p))$, first, find $3g(p)$

$g(x) = log_3 x \rightarrow g(p) = log_3 p \rightarrow 3g(p) = 3log_3 p = log_3 p^3$

Now, find $f(3g(p))$: $f(x) = 3^x \rightarrow f(log_3 p^3) = 3^{log_3 p^3}$

Logarithms and exponentials with the same base cancel each other. This is true because logarithms and exponentials are inverse operations. Then: $f(log_3 p^3) = 3^{log_3 p^3} = p^3$

59) Answer: B.

$$x_1 = \frac{6y + \frac{r}{r+4}}{\frac{z}{4}} = \frac{6y + \frac{r}{r+4}}{\frac{4 \times 10}{z}} = \frac{6y + \frac{r}{r+4}}{4 \times \frac{10}{z}} = \frac{1}{4} \times \frac{6y + \frac{r}{r+4}}{\frac{10}{z}} = \frac{x}{4}$$

60) Answer: D.

The amplitude in the graph of the equation $y = asinbx$ is a. (a and b are constant)

In the equation $y = sinx$, the amplitude is 1 and the period of the graph is 2π.

The only option that has four times the amplitude of graph $y = sin\,x$ is $y = 4 + 4sin\,2x$ for the half period $sin2x = sin2\pi \Rightarrow 2x = 2\pi \Rightarrow x = \pi$

They both have the amplitude of 4 and period of π.

ACT Mathematics

Practice Tests 2

1) Answer: D.

$3,260,000 = 3.26 \times 10^6$

2) Answer: A.

$(x^4)^{\frac{5}{8}} = x^{4 \times \frac{5}{8}} = x^{\frac{20}{8}} = x^{\frac{5}{2}}$

3) Answer: A.

To simplify the fraction, multiply both numerator and denominator by i.

$\dfrac{2-5i}{-4i} \times \dfrac{i}{i} = \dfrac{2i-5i^2}{-4i^2}$

$i^2 = -1$, Then: $\dfrac{2i-5i^2}{-4i^2} = \dfrac{2i-5(-1)}{-4(-1)} = \dfrac{2i+5}{4} = \dfrac{i}{2} + \dfrac{5}{4}$

4) Answer: B.

$256 = 4^4 \quad \rightarrow 4^x = 4^4 \rightarrow x = 4$

5) Answer: D.

Here is the list of all prime numbers between 1 and 20:

2, 3, 5, 7, 11, 13, 17, 19

The sum of all prime numbers between 1 and 20 is:

$2 + 3 + 5 + 7 + 11 + 13 + 17 + 19 = 77$

6) Answer: E.

Solve for x. $\sqrt[3]{6x} = \sqrt[3]{y}$

Cube both sides of the equation: $(\sqrt[3]{6x})^3 = (\sqrt[3]{y})^3 \rightarrow 6x = y \rightarrow x = \dfrac{y}{6}$

7) Answer: B.

$\text{average} = \dfrac{\text{sum of terms}}{\text{number of terms}}$

The sum of the weight of all girls is: $18 \times 45 = 810 \text{ kg}$

The sum of the weight of all boys is: $22 \times 50 = 1,100 \text{ kg}$

The sum of the weight of all students is: $810 + 1,100 = 1,910 \text{ kg}$

$\text{average} = \dfrac{1,910}{40} = 47.75$

8) **Answer: E.**

$$y = (-3x^6)^3 = (-3)^3(x^6)^3 = -27x^{18}$$

9) **Answer: C.**

Plug in the value of x and y. $x = 5$ and $y = -3$

$$5(x + y) + (7 - x)^3 = 5(5 + (-3)) + (7 - 5)^3 = 5(2) + (2)^3 = 10 + 8 = 18$$

10) **Answer: B.**

The question is this: 246.50 is what percent of 725?

Use percent formula: part $= \frac{\text{percent}}{100} \times$ whole

$$246.50 = \frac{\text{percent}}{100} \times 725 \Rightarrow 246.50 = \frac{\text{percent} \times 725}{100} \Rightarrow 24{,}650 = \text{percent} \times 725$$

$$\Rightarrow \text{percent} = \frac{24{,}650}{725} = 34$$

246.50 is 34% of 725. Therefore, the discount is: $100\% - 34\% = 66\%$

11) **Answer: A.**

$$\left(\frac{f}{g}\right)(x) = \frac{f(x)}{g(x)} = \frac{2x - 9}{4x^2 - 3x}$$

12) **Answer: A.**

The equation of a line is: $y = mx + b$, where m is the slope and b is the y-intercept.

First find the slope: $m = \frac{y_2 - y_1}{x_2 - x_1} = \frac{7 - (-1)}{4 - 2} = \frac{8}{2} = 4$

Then, we have: $y = 4x + b$

Choose one point and plug in the values of x and y in the equation to solve for b.

Let's choose the point $(2, -1)$

$$y = 4x + b \rightarrow -1 = 4(2) + b \rightarrow -1 = 8 + b \rightarrow b = -9$$

The equation of the line is: $y = 4x - 9$

13) **Answer: B.**

Use simple interest formula: $I = prt$ (I = interest, p = principal, r = rate, t = time)

$$I = (24{,}000)(0.0225)(4) = 2{,}160$$

14) **Answer: E.**

Number of visiting fans: $\frac{7 \times 18{,}000}{9} = 14{,}000$

15) Answer: E.

The sum of all angles in a quadrilateral is 360 degrees.

Let x be the smallest angle in the quadrilateral. Then the angles are: $3x, 7x, 9x, 11x$

$3x + 7x + 9x + 11x = 360 \to 30x = 360 \to x = 12$

The angles in the quadrilateral are $36°$, $84°$, $108°$, and $132°$

16) Answer: B.

$4sin^2a + 4cos^2a = 4(sin^2a + cos^2a) = 4(1) = 4$, then:

$x + 4 = 10 \to x = 6$

17) Answer: C.

Formula for the area of a circle is: $A = \pi r^2$

Using 121 for the area of the circle we have: $121 = \pi r^2$

Let's solve for the radius (r).

$\frac{121}{\pi} = r^2 \to r = \sqrt{\frac{121}{\pi}} = \frac{11}{\sqrt{\pi}} = \frac{11}{\sqrt{\pi}} \times \frac{\sqrt{\pi}}{\sqrt{\pi}} = \frac{11\sqrt{\pi}}{\pi} \to d = 2r = 2 \times \frac{11\sqrt{\pi}}{\pi} \to d = \frac{22\sqrt{\pi}}{\pi}$

18) Answer: C.

Length of the rectangle is: $\frac{5}{7} \times 70 = 50$

perimeter of rectangle is: $2 \times (50 + 70) = 240$

19) Answer: E.

The angle x and 65 are supplement angles. Therefore:

$x + 65 = 180$

$180° - 65° = 115°$

20) Answer: A.

Solving Systems of Equations by Elimination

Multiply the first equation by (-3), then add it to the second equation.

$\begin{matrix} -3(x + 3y = 14) \\ 3x + 2y = 21 \end{matrix} \Rightarrow \begin{matrix} -3x - 9y = -42 \\ 3x + 2y = 21 \end{matrix} \Rightarrow -7y = -21 \Rightarrow y = 3$

Plug in the value of y into one of the equations and solve for x.

$x + 3(3) = 14 \Rightarrow x + 9 = 14 \Rightarrow x = 14 - 9 \Rightarrow x = 5$

21) Answer: A.

The sum of supplement angles is 180. Let x be that angle. Therefore, $x + 11x = 180$

$\Rightarrow 12x = 180$, divide both sides by 12: $x = 15$

22) Answer: D.

Six times of 12,000 is 72,000. One ninth of them cancelled their tickets.

One ninth of 72,000 equal 8,000 ($\frac{1}{9} \times 72,000 = 8,000$).

64,000 (72,000 – 8,000 = 64,000) fans are attending this week

23) Answer: E.

$sin\alpha = \frac{\sqrt{2}}{2} \Rightarrow$ Since $sin\alpha = \frac{opposite}{hypotenuse}$, we have the following right triangle. Then,

$c = \sqrt{2^2 - (\sqrt{2})^2} = \sqrt{4 - 2} = \sqrt{2}$

$cos\alpha = \frac{\sqrt{2}}{2}$

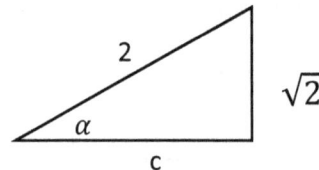

24) Answer: D.

In any squares measure of all angles equals 360°.

25) Answer: B.

The equation of a circle in standard form is:

$(x - h)^2 + (y - k)^2 = r^2$, where r is the radius of the circle.

In this circle the radius is 4. $r^2 = 16 \rightarrow r = 4$

$(x + 4)^2 + (y - 3)^2 = 4^2$

Area of a circle: $A = \pi r^2 = \pi(4)^2 = 16\pi$

26) Answer: C.

the population is increased by 18% and 20%. 18% increase changes the population to 118% of original population.

For the second increase, multiply the result by 120%.

$(1.18) \times (1.20) = 1.416 = 141.6\%$

41.6 percent of the population is increased after two years.

27) **Answer: D.**

$5x^6y^2 + 4x^3y^5 - (3x^6y^2 - 6x^3y^5) = 5x^6y^2 + 4x^3y^5 - 3x^6y^2 + 6x^3y^5 =$

$2x^6y^2 + 10x^3y^5$

28) **Answer: E.**

Frist factor the function: $f(x) = 4x^3 + 40x^2 + 84x = 4x(x + 3)(x + 7)$

To find the zeros, $f(x)$ should be zero. $f(x) = 4x(x + 3)(x + 7) = 0$

Therefore, the zeros are: $x = 0$

$(x + 3) = 0 \Rightarrow x = -3 \,;\, (x + 7) = 0 \Rightarrow x = -7$

29) **Answer: A.**

The relationship among all sides of right triangle $30°, \ 60°, \ 90°$ is provided in the

following triangle:

Sine of 30° equals to: $\dfrac{opposite}{hypotenuse} = \dfrac{x}{2x} = \dfrac{1}{2}$

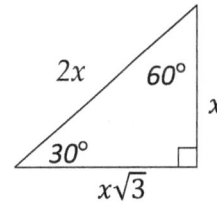

30) **Answer: D.**

Let x be the length of AB, then: $48 = \dfrac{x \times 8}{2} \to x = 12$

The length of AC $= \sqrt{12^2 + 16^2} = \sqrt{400} = 20$

The perimeter of $\Delta ABC = 12 + 20 + 16 = 48$

31) **Answer: C.**

$x_{1,2} = \dfrac{-b \pm \sqrt{b^2 - 4ac}}{2a}$

$ax^2 + bx + c = 0 \Rightarrow 4x^2 + 7x - 2 = 0$, then: a = 4, b = 7 and c = − 2

$x = \dfrac{-7 + \sqrt{7^2 - 4 \times 4 \times (-2)}}{2 \times 4} = \dfrac{1}{4} \,;\, x = \dfrac{-7 - \sqrt{7^2 - 4 \times 4 \times (-2)}}{2 \times 4} = -2$

32) **Answer: A.**

Let x be the cost of one-kilogram orange, then:

$5x + (4 \times 3.2) = 72.8 \to 5x + 12.8 = 72.8 \to 5x = 72.8 - 12.8 \to 5x = 60 \to$

$x = \dfrac{60}{5} = \$12$

33) Answer: B.

Simplify the expression.

$$\sqrt{\frac{x^2}{7} + \frac{2x^2}{49}} = \sqrt{\frac{7x^2}{49} + \frac{2x^2}{49}} = \sqrt{\frac{9x^2}{49}} = \sqrt{\frac{9}{49}x^2} = \sqrt{\frac{9}{49}} \times \sqrt{x^2} = \frac{3}{7} \times x = \frac{3x}{7}$$

34) Answer: C.

Write the ratio of $16a$ to $12b$, $\frac{16a}{12b} = \frac{1}{20}$

Use cross multiplication and then simplify.

$$16a \times 20 = 12b \times 1 \rightarrow 320a = 12b \rightarrow a = \frac{12b}{320} = \frac{3b}{80}$$

Now, find the ratio of a to b.

$$\frac{a}{b} = \frac{\frac{3b}{80}}{b} \rightarrow \frac{3b}{80} \div b = \frac{3b}{80} \times \frac{1}{b} = \frac{3b}{80b} = \frac{3}{80}$$

35) Answer: E.

Based on triangle similarity theorem:

$$\frac{a}{a+b} = \frac{c}{6} \rightarrow c = \frac{6a}{a+b} = \frac{12\sqrt{2}}{2\sqrt{2}+4\sqrt{2}} = 2$$

$\rightarrow$ area of shaded region is: $\left(\frac{c+6}{2}\right)(b) = \frac{8}{2} \times 4\sqrt{2} = 16\sqrt{2}$

36) Answer: D.

First find the number of pants sold in each month.

January: 122, February: 114, March: 120, April: 63, May: 84, June: 72

Check each option provided.

A. There is a decrease from January to February.

B. February and March,

$$\left(\frac{120-114}{120}\right) \times 100 = \frac{6}{120} \times 100 = 5\%$$

C. There is a decrease from March to April.

D. April and May: there is an increase from April to May.

$$\left(\frac{84-63}{84}\right) \times 100 = \frac{21}{84} \times 100 = 25\%$$

E. There is a decrease from May to June.

37) Answer: C.

First, order the number of shirts sold each month: $130, 140, 144, 150, 160, 170$.

mean is: $\dfrac{130+140+144+150+160+170}{6} = \dfrac{894}{6} = 149$

Put the number of shoes sold per month in order:

$25, 25, 25, 35, 35, 40$; median is: $\dfrac{25+35}{2} = 30$

38) Answer: B.

The ratio of number of pants to number of shoes in March equals $\dfrac{120}{40}$.

Three-twelfth of this ratio is $\left(\dfrac{3}{12}\right)\left(\dfrac{120}{40}\right)$. Now, let x be the number of shoes needed to

be added in February.

$\dfrac{114}{35+x} = \left(\dfrac{3}{12}\right)\left(\dfrac{120}{40}\right) \rightarrow \dfrac{114}{35+x} = \dfrac{360}{480} = \dfrac{3}{4} \rightarrow 456 = 3(35+x) \rightarrow 456 = 105 + 3x \rightarrow$

$3x = 351 \rightarrow x = 117$

39) Answer: B.

The probability of choosing a spades or diamonds is $\dfrac{26}{52} = \dfrac{1}{2}$

40) Answer: A.

Let x be the number of shoes the team can purchase. Therefore, the team can purchase

$120\,x$.

The team had \$34,000 and spent \$21,000. Now the team can spend on new shoes

\$13,000 at most.

Now, write the inequality: $120x + 21,000 \le 34,000$

41) Answer: A.

Plug in the value of x in the equation and solve for y.

$3y = \dfrac{7x^4}{16} + 14 \rightarrow 3y = \dfrac{7(2)^4}{16} + 14 \rightarrow 3y = \dfrac{7(16)}{16} + 14 \rightarrow 3y = 7 + 14 = 21$

$\rightarrow 3y = 21 \rightarrow y = 7$

42) Answer: E.

Use formula of rectangle prism volume.

$V = \text{(length) (width) (height)} \Rightarrow 6,000 = (6)(20)(\text{height}) \Rightarrow \text{height} = 6,000 \div 120 = 50$

43)Answer: E.

The ratio of boy to girls is 9:11. Therefore, there are 9 boys out of 20 students. To find the answer, first divide the total number of students by 20, then multiply the result by 9. $480 \div 20 = 24 \Rightarrow 24 \times 9 = 216$

44)Answer: C.

Plug in the value of each option in the inequality.

A. 2 $(2-5)^2 + 2 > 4(2) + 5 \to 11 > 13$ No!

B. 6 $(6-5)^2 + 2 > 4(6) + 5 \to 3 > 29$ No!

C. 0 $(0-5)^2 + 2 > 4(0) + 5 \to 27 > 5$ Bingo!

D. 5 $(5-5)^2 + 2 > 4(5) + 5 \to 2 > 25$ No!

E. 4 $(4-5)^2 + 2 > 4(4) + 5 \to 3 > 21$ No!

45)Answer: A.

$(x+5)(x+p) = x^2 + (5+p)x + 5p \to 5 + p = 7 \to p = 2 \text{ and } r = 5p = 10$

46)Answer: C.

First, find the number.

Let x be the number. Write the equation and solve for x.

160% of a number is 72, then:

$1.6 \times x = 72 \Rightarrow x = 72 \div 1.6 = 45$

40% of 45 is: $0.4 \times 45 = 18$

47)Answer: D.

If the length of the box is 45, then the width of the box is third of it, 15, and the height of the box is 5 (third of the width). The volume of the box is:

V = (length) × (width) × (height) = (45) × (15) × (5) = 3,375

48)Answer: C.

Let x be the smallest number. Then, these are the numbers:

$x, x+1, x+2, x+3, x+4, x+5$

$\text{average} = \frac{\text{sum of terms}}{\text{number of terms}} \Rightarrow 35 = \frac{x+(x+1)+(x+2)+(x+3)+(x+4)+(x+5)}{6} \Rightarrow 35 = \frac{6x+15}{6}$

$\Rightarrow 210 = 6x + 15 \Rightarrow 195 = 6x \Rightarrow x = 32.5$

49) Answer: D.

Formula for the Surface area of a cylinder is:

$$SA = 2\pi r^2 + 2\pi rh \rightarrow 112\pi = 2\pi r^2 + 2\pi r(10) \rightarrow r^2 + 10r - 56 = 0$$

Factorize and solve for r.

$$(r + 14)(r - 4) = 0 \rightarrow r = 4 \quad or \quad r = -14 \; (unacceptable)$$

50) Answer: C.

Since the triangle ABC is reflected over the y-axis, then all values of y's of the points don't change and the sign of all x's change.

(remember that when a point is reflected over the y-axis, the value of y does not change and when a point is reflected over the x-axis, the value of x does not change).

Therefore:

$(7, -2)$ changes to $(-7, -2)$

$(-5, -3)$ changes to $(5, -3)$

$(4, 6)$ changes to $(-4, 6)$

51) Answer: B.

The equation of a line in slope intercept form is: $y = mx + b$

Solve for y.

$$14x - 2y = 8 \Rightarrow -2y = 8 - 14x \Rightarrow y = (8 - 14x) \div (-2) \Rightarrow$$

$y = 7x - 4 \rightarrow$ The slope is 7.

The slope of the line perpendicular to this line is:

$$m_1 \times m_2 = -1 \Rightarrow 7 \times m_2 = -1 \Rightarrow m_2 = -\frac{1}{7}$$

52) Answer: B.

The area of rectangle is: $7 \times 5 = 35 \; cm^2$

The area of circle is: $\pi r^2 = \pi \times (\frac{10}{2})^2 = 3 \times 25 = 75 \; cm^2$

Difference of areas is: $75 - 35 = 40$

53) Answer: E.

$$f(g(x)) = 5 \times (\frac{3}{x})^2 + 2 = \frac{45}{x^2} + 2$$

54) Answer: D.

Use the information provided in the question to draw, the shape. Use Pythagorean Theorem: $a^2 + b^2 = c^2$

$18^2 + 24^2 = c^2 \Rightarrow 324 + 576 = c^2$

$\Rightarrow 900 = c^2 \Rightarrow c = 30$

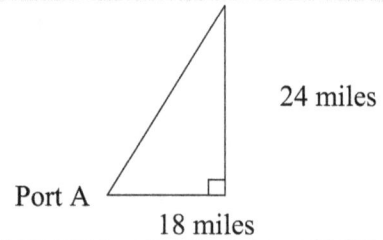

55) Answer: C.

Let L be the length of the rectangular and W be the width of the rectangular.

Then, $L = 5W + 3$

The perimeter of the rectangle is 114 meters. Therefore:

$$2L + 2W = 114$$

$$L + W = 57$$

Replace the value of L from the first equation into the second equation and solve for W:

$$(5W + 3) + W = 57 \rightarrow 6W + 3 = 57 \rightarrow 6W = 54 \rightarrow W = 9$$

The width of the rectangle is 9 meters, and its length is: $L = 5W + 3 = 5(9) + 3 = 48$

The area of the rectangle is: length × width = $48 \times 9 = 432$

56) Answer: B.

Let x be the number of adult tickets and y be the number of student tickets. Then:

$x + y = 22$

$9.50x + 6.50y = 167$

Use elimination method to solve this system of equation. Multiply the first equation by -6.5 and add it to the second equation.

$-6.5(x + y = 22) \Rightarrow -6.5x - 6.5y = -143$

$9.50x + 6.50y = 167 \Rightarrow 3x = 24 \rightarrow x = 8$

There are 8 adults' tickets and 14 student tickets.

57) Answer: A.

$x - 6 \geq 10 \rightarrow x \geq 10 + 6 \rightarrow x \geq 16$

Or $x - 6 \leq -10 \rightarrow x \leq -10 + 6 \rightarrow x \leq -4$

Then, solution is: $x \geq 16 \cup x \leq -4$

58) Answer: B.

$\tan = \frac{opposite}{adjacent}$, and $\tan x = \frac{10}{24}$, therefore, the opposite side of the angle x is 10 and

the adjacent side is 24. Let's draw the triangle.

Using Pythagorean theorem, we have:

$a^2 + b^2 = c^2 \to 10^2 + 24^2 = c^2 \to 100 + 576 = c^2 \to c = 26$

$\sin x = \frac{opposite}{hypotenuse} = \frac{10}{26} = \frac{5}{13}$

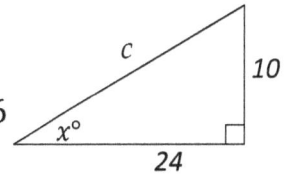

59) Answer: E.

First, find $2A$.

$A = \begin{bmatrix} 2 & 3 \\ -1 & 4 \end{bmatrix} \Rightarrow 2A = 2 \times \begin{bmatrix} 2 & 3 \\ -1 & 4 \end{bmatrix} = \begin{bmatrix} 4 & 6 \\ -2 & 8 \end{bmatrix}$

Now, solve for $2A - B$:

$\begin{bmatrix} 4 & 6 \\ -2 & 8 \end{bmatrix} - \begin{bmatrix} 5 & 1 \\ -1 & 4 \end{bmatrix} = \begin{bmatrix} 4-5 & 6-1 \\ -2-(-1) & 8-4 \end{bmatrix} = \begin{bmatrix} -1 & 5 \\ -1 & 4 \end{bmatrix}$

60) Answer: B.

The amplitude in the graph of the equation $y = a\cos bx$ is a. (a and b are constant)

In the equation $y - 3 = 6\cos 4x$, the amplitude is 6.

"End"

www.ingramcontent.com/pod-product-compliance
Lightning Source LLC
Chambersburg PA
CBHW081325090426

42737CB00017B/3037